DAVID A. WATTERS

Handbook for Writing Technical Proposals That Win Contracts

Donald V. Helgeson has spent more than 20 years in the
aerospace industry in systems engineering, marketing,
and program management. During this time he has written
and managed technical proposals for government contracts—
including several in the $100 million range for the Army,
Air Force, and NASA. He has also worked on proposals
for the Navy, DARPA, the Jet Propulsion Lab, and various
government agencies. He is a retired army Lieutenant
Colonel and now works as a consultant to large aerospace
firms on marketing policy and proposal preparation.

HANDBOOK FOR WRITING
TECHNICAL PROPOSALS THAT WIN CONTRACTS

Donald V. Helgeson

DAVID A. WATTERS

PRENTICE-HALL, INC.
Business & Professional Division
Englewood Cliffs, New Jersey

First Printing . . . August 1985

Editor: George E. Parker

Library of Congress Cataloging in Publication Data
Helgeson, Donald V.
    Handbook for writing technical proposals that
win contracts.

    Includes index.
    1. Technical writing.   2. Proposal writing in
research.   3. Research and development contracts.
I. Title.
T11.H45 1985   808'.066658   85-3549

ISBN 0-13-379686-8

Prentice-Hall International, Inc., *London*
Prentice-Hall of Australia, Pty. Ltd., *Sydney*
Prentice-Hall Canada Inc., *Toronto*
Prentice-Hall of India Private Ltd., *New Delhi*
Prentice-Hall of Japan, Inc., *Tokyo*
Prentice-Hall of Southeast Asia Pte. Ltd., *Singapore*
Whitehall Books, Ltd., *Wellington, New Zealand*
Editora Prentice-Hall do Brasil Ltda., *Rio de Janeiro*
Prentice-Hall Hispanoamericana, S.A., *Mexico*

This book is dedicated to all the victims of mismanaged proposals—past, present, and future; and to the engineers and technicians who stoically toil in the trenches, for they are the true heroes of most business success stories in the world of competitive bidding.

Acknowledgment is hereby tendered to all the poor
managers I have endured, because they taught me
so well how *not* to do things. . . .

And to the good ones . . .

God bless you—both of you!

Contents

ILLUSTRATIONS

Figures

ABBREVIATIONS USED IN THIS BOOK

ADP   Automatic data processing
BAFO   Best and final offer
B&P   Bid and proposal (funds)
CBD   Commerce Business Daily
CCTV   Closed circuit television
CDL   Contract data list
CDRL   Contract data requirements list
CEO   Chief executive officer
CPAF   Cost plus award fee
CPFF   Cost plus fixed fee
DAR   Defense acquisition regulations
DID   Data item description
DOD   Department of Defense
EEO   Equal employment opportunity
E-Tech   Electronic technician
G&A   General and administrative (costs)
GFP   Government furnished property
G-Z   Military intelligence (information on competitors)
IAW   In accordance with
IFB   Invitation for bid
IGCE   Independent government cost estimate
IQ   Intelligence quotient
K   To engineers, K represents thousand (kilo)
M   To engineers, M represents million (mega)
MIS   Management information system
NASA   National Aviation and Space Agency
ODC   Other direct charges (in costing)
OMB   Office of Management and Budget
O&M   Operation and maintenance
PCO   Purchasing and contracting officer
PERT   Program Evaluation and Review Technique

POL   Petroleum, oil, and lubricants

PR   Public relations

Q&A   Question and answer (session or conference)

QC   Quality control

R&D   Research and development

RFP   Request for proposal

SBA   Small Business Administration

SEATA   Systems Engineering and Technical Assistance (contractor)

SOP   *Standing* operating procedure (not *standard*)

SOW   Statement of Work of Scope of Work

SSA   Source Selection Authority

SSAC   Source Selection Advisory Committee

SSEB   Source Selection Evaluation Board

TDY   Temporary duty

TM   Telemetry

VIP   Very important person

Introduction:
How This Book Will Help You
Write Technical Proposals

Twenty years ago the ABC Company had about 200 employees, and the XYZ Company likewise had about 200 employees. Each of these companies was totally dedicated to performing technical services primarily (though not exclusively) for the Government. Today the ABC Company has over 15,000 employees and does over $600 million of business. The XYZ Company has about 1200 employees and a dwindling gross of about $40 million of business. Why this remarkable difference in performance by two similarly organized companies?

The answer is QUALITY—the quality of management, the quality of employees, and the quality of their respective marketing capabilities. Success in the marketing of engineering and scientific services depends on the quality of technical proposals, for the written word prepared in response to the customer's needs is the only legitimate basis (other than price, of course) that can be considered in awarding competitive contracts. Of course, every company that aspires to grow and prosper in the engineering and scientific services business has a carefully developed *modus operandi* for proposal development the very life blood of their business. Right? Wrong. It is truly astounding that top management generally takes a rather cavalier attitude toward this vital factor in assuring future growth. Perhaps it's because it is more fun to wine and dine the customer and delude oneself into thinking a contract can be won with charm and gratuities. Perhaps it's because some managers regard proposal preparation as a tedious bore and a necessary evil that can be delegated to engineers from the field on an *ad hoc* basis and then forgotten. Whatever the reason, the actual preparation of the proposal is the weakest link in the marketing chain in most companies. The preparation of winning proposals is an art and a science requiring research, analysis, communications skill, organization, and imagination. And the company that fails to recognize the essentiality of a carefully planned and professionally prepared proposal is headed for zero growth.

This book is designed to strengthen that "weak link" by showing proposal contributors and proposal managers how to write a winning proposal. It is not concerned with peripheral details such as printing, editing, graphic arts, and customer relations. Nor is it concerned with the cost proposal. These tasks are performed by specialists normally made available to the proposal manager by management and who are presumably competent and experienced.

The first part of the book is concerned with the preliminary preparations for and, finally, the actual writing of the proposal. You will see how the typical customer-requirements document is organized and what to look for, and then how to analyze—dissect this document. This is a very important step in the proposal writing process. Then comes the actual writing phase: how to organize the proposal; how to approach the various sections; how to get started; some guidelines to writing style.

The next chapter is designed to dispel some of the misconceptions often held by proposal writers as to how the customer actually evaluates your proposal. (They don't throw darts; they don't weigh them.) As a matter of fact, the customer generally has a very well organized, more or less standard procedure for evaluating your proposal and for documenting his evaluation so as to justify the award (to higher headquarters) and respond to a protest or (worst case) a law suit. It is mandatory that a proposal writer study this section carefully in order that he understand the thought process the evaluators go through. In this way the proposal writer can intelligently evaluate his own proposal on the basis of what he knows the customer wants to see.

The last two chapters of this book are directed toward one of the most challenging and demanding jobs to be found in the business world. That is the organizing and managing of a proposal team and providing the guidance and leadership that create a successful proposal. What is presented here is the distillation of many years of experience down in the trenches; you will find it eminently practical and realistic—I'm telling it like it is in the real world.

You must bear in mind that a proposal is, in the last analysis, a sales document. But it is more important that you also remember who your customer is, who you are addressing and what are the customer's objectives. You are addressing professional-level people, and the high-pressure sales pitch to them is a turn-off, counterproductive. Surprisingly, most of the "expert consultants" I have encountered in this business emphasize this sales pitch approach to proposal preparation. They should know better. The poor beleaguered engineer or technician or administrator who has been dragooned into evaluating your proposal couldn't care less about all your past accomplishments or vast resources. He has a set procedure to follow (as explained herein) and doesn't want to read any extraneous, irrelevant nonsense (sales pitches). That excess verbiage only gets in the way as he performs the job of evaluating the meat in your proposal. Your reader wants to see responses to the basic document, facts, logic, solutions. Even if he was beguiled by all your bragging and cleverness, he is not in a position to influence the contract award on that basis. The only place where the sales pitch garbage belongs is in the executive summary or the cover letter. Why? Because only the top-level people who read the executive summary are in a position to influence the award based on such subjective criteria. (It is sometimes true there is a place in your technical proposal for weaving in a subtle message, and that will be elaborated in the chapter on proposal management.)

I once had an unpleasant experience that illustrates the point. We were in the initial stages of writing a proposal for a very complex operation involving over 400 man-years per year, and comprising the operation and maintenance of no less than 50 discrete hardware/software systems that had to be integrated into one sophisticated operation. When the Request for Proposal (RFP) came out, I made an outline for the entire proposal to provide some 40 people with a road map to guide them through this labyrinthian document that would total over 700 pages. Then along came the *company president*, who had just taken a "Gee Whiz" seminar in proposal writing, apparently presented by a bunch of used car salesmen. He barely glanced at this outline, which I had painstakingly produced through four days of hard labor, and then gave me a lecture on getting "themes" (translated "sales pitch") into the proposal. "Gotta hit our competitions' weaknesses and emphasize our strong points," he pontificated. "Sure, that's fine," I agreed, "but we can't clutter up an *outline* with that stuff. With the engineers and technicians we have available (with no proposal writing experience or training) this sort of thing in an outline will just confuse them and distract them from their primary objective. And that is to write coherent, simple declarative sentences that adequately respond to the RFP." Fortunately for all concerned, he had to leave town the next day, so I could ignore his advice with impunity. I made out a list of themes and incorporated them into the executive summary where such things belong, and we won the proposal in spite of the gratuitous assistance of this misguided individual.

It is just such futile, aimless, ill advised tinkering as this that causes proposal preparation in most companies to be an almost insufferable ordeal for the engineers and scientists who have to write them. Because of all the waste motion resulting from poor planning, lack of organization, failure to develop a sensible methodology, and amateurish interference from management, the proposal effort usually ends in sheer, utter panic and chaos. I once ended up spending New Year's Eve in a frantic attempt to rescue a proposal for a recompete of a contract where the company had been the incumbent for ten years. Even though the incumbent personnel had already had four months to write the proposal, here we were in the usual panic situation and for the usual reasons.

At the time, one of the company vice presidents asked me rhetorically, "Don, why does every proposal we write have to be like it was the first one we ever did?" (I used the word "rhetorically" because, predictably, he didn't pay any attention to my answer.) My answer is that proposal writing *doesn't have* to be that way. If managers would apply the procedures set out in this book and provide the proper training to their people such as is set forth in this book, there would not be all this waste motion, disorganization, confusion, and finally, panic—followed by disaster.

While I am on the subject, I would like to say a few words about management—leadership in general. After all, this is a book about management, too. It has been my observation that we have no scarcities in the United States except one—good leadership. We have been blessed with natural resources, a skilled work force, a great educational system—at least at the college level. So, then, why are the Japanese beating hell out of us in the auto industry where we had a 50-year head start?

Japan (the size of Montana with virtually no natural resources), rising from the ruins of a war fought less than 50 years ago is putting our workers out on the street. The answer is leadership—management. We should start learning from the Japanese experience, both labor leaders and business managers. Most managers I have known equate their position in the company with their IQ. "I am an executive vice president, ergo I know more about *anything* than *anybody* except the president." The really great managers I have known are continually picking the brains of subordinates, trying to get as smart as they are. And, they are not above putting themselves in a subordinate position on a project in order to achieve the teamwork necessary to get the project completed successfully. For more details on Japanese management, talk to the heads of the auto workers' or steel workers' unions.

This book is addressed primarily to the art of preparing proposals for engineering development, technical services, and research and development projects. The principles involved are the same, whether it be management services or computer hardware, and whether it is to the Government or to industry. Only the terminology differs.

This book is dedicated especially to the long-suffering and often maligned engineers who are suddenly displaced from their normal milieu: the solving of engineering problems and implementation of engineering solutions. Engineers who work in the field, providing technical or logistic support services are accustomed to dealing with day-to-day practical realities. The problem is usually well defined, susceptible to quantification in more or less precise terms, and amenable to a pragmatic, logical solution. Proposal writing, on the other hand, involves sifting through reams of legalistic boiler plate, identifying what needs to be addressed, and then writing what is essentially a sales document. This, most assuredly, is not the *milieu* of the engineer.

The fact remains that engineers and scientists are the only ones with the technical background to write a technical proposal. The best technical writers in the world cannot begin to take their place. The problem is that no one takes the trouble to provide engineers with some orientation and training in the art of proposal writing. No wonder they are resentful, uncomfortable, intimidated, and sometimes (rightfully so) outraged when assigned to proposal preparation. As a matter of fact, in most companies there is no one around who is capable of giving them the proper orientation and training and provide them with the concepts, the tools of the trade, for writing a proposal.

That is what this book intends to do—to give the proposal contributor, the proposal manager, too, the concepts, the orientation, the *modus operandi*, for writing a successful technical proposal. If the lessons in this book are absorbed and the proposal manager adopts a systematic method for proposal preparation as described herein, everyone will find that there will be a great reduction of waste motion, elimination of frustrations, and perhaps some of you will even find that proposal preparation can be fun.

Note: For purposes of clarity, I felt that the constant use of he/she and him/her would interrupt the attention of the reader rather than demonstrate an attempt to eliminate sexism. Therefore, any time a reference is made to "he" or "him," please read these terms as generic abbreviations for he/she and him/her.

CHAPTER ONE

WHERE YOUR HOMEWORK BEGINS

The successful proposal has its origins in a process begun months, even years, before the first announcement appears in the media or the CBD (Commerce Business Daily)—or before the Request for Proposal (RFP) comes out and before the first word is set on paper. As in most momentous projects, the proposal preparation begins with research. Research begins with the marketing people in your company, the "intelligence" people, the collectors of information, the politicians, the image makers, the molders of opinion, the creators of biases—biases in favor of your company and against your competitors. And that is where Chapter 1 begins.

Here's to success—and the defeat of your enemies.

ESTABLISHING GOOD CUSTOMER RELATIONS

The first step in the acquisition of any new business is the establishment of good customer relations. This is the primary function of the professional marketing people. Various companies use various euphemisms to identify their marketeers. Some companies call them program development, others, advanced programs or business planning. Whatever they are called, these people are the up-front people who are "tracking" business opportunities and potential contracts well in advance of any public announcement of an impending RFP. They are an indispensable part of any organization that has aspirations for growth. They are also a special breed of people. They must have an engineering or scientific background, not necessarily in depth but enough to know how to use the buzz words and understand and articulate scientific-based concepts. They must also have a certain amount of charm, poise, initiative, and—shall I say—impudence to knock on doors or put a foot in doors on occasion and, finally, the most important, the good judgment of knowing when and when not to exercise those talents.

How do they establish good customer relations? First, by projecting a good image of your company. What constitutes a good image will, of course, vary with the eye of the beholder. What appears to be good old conviviality to one customer may come across as irresponsible playboy stuff to another. The successful marketeer must have the intuition and empathy to discern the image that the customer wants to see as well as the flexibility to respond to these wants.

Generally speaking, the marketeers should project an image of confidence, knowledge of their customers' environments and of their needs, a very thorough knowledge of their own company and of its capabilities, and finally a desire to provide customers with any information they need to fulfill their requirements. In short, they must know their customers, must know their products, and must be adept at showing how their companies can help their customers. The marketeers, in the context of this book, are dealing with top echelon management or engineering people. They must meet them on their own ground, in their arena. Their weapons are the slide projector, the flip chart, and all the technical and communication skills they can muster. Note that I did not say anything about persuasiveness. That's because facts, logic, efficiency, and cost effectiveness are the persuaders in this arena.

In addition to establishing good customer relations, the marketing people gather all pertinent information possible in regard to the target contract. This will include background information, organization charts, political situation, incumbent weaknesses and strengths, potential teaming arrangements, in-depth knowledge of the contract requirements, personalities, personal preferences of the customer's key personnel, and so on *ad infinitum*. Based on this information, the marketing people help devise the bidding strategy and also help in preparation of the executive summary. But most important of all, this is the information that is vital to company management in making the Bid/No Bid decision.

Of course, if you are the incumbent, this function is primarily the responsibility of the program manager and all his people through their performance on the contract.

BID/NO BID CRITERIA

The marketing people play a crucial role in making the Bid/No Bid decision. They are usually the only ones who have the information upon which to base this decision. Therefore, they are the ones who must convince management of the viability of any particular bid opportunity. The Bid/No Bid decision-making process is of vital importance to the survival of small companies and to the stockholders of large companies. Preparing a bid proposal on a large contract is an expensive game. The proposal costs vary directly with the size of the contract. The B&P (bid and proposal) budget on large contracts often runs from $100 thousand to $1 million. In spite of this great outlay of cash, most companies seem to approach this critical decision in a haphazard, euphoric, unsystematic manner, that is, without any discernible criteria upon which to base their decision. Amazingly, some companies shotgun everything in sight. They usually end up in bankruptcy. Others play a pat hand, looking for the sure thing. They stagnate. (There are no sure things.) Others naively rush in where angels fear to tread. I know one company that rushed headlong into the Middle East cauldron to get "some of that easy oil money from the Arabs." I doubt if anyone in the company could even find Kuwait on the map. After spending hundreds of thousands of dollars that they could ill afford, they finally came to the realization that doing business with the Arabs was a whole new ball game, and heads began to roll in the executive suite.

In order to help business executives to keep their heads, I have devised a set of criteria to apply in making Bid/No Bid decisions. If you are not an executive, you might show the following to your boss. It might get you on the Christmas bonus list.

I strongly advise all marketing managers/vice presidents or business managers or whoever must make these decisions to paste a copy of this on their office walls, or in their desks, or on their bathroom mirrors. It might save them from making a series of bum decisions and wasting a lot of the corporation's money. "The job you save may be your own!"

Summary: Bid/No Bid Criteria

1. Incumbent's performance ratings—past 5 years

2. Incumbent's program manager
   a. Qualifications—General reputation
   b. Performance history on the program
   c. Predecessor program managers, if any

3. Procurement office history on recompetitions

4. Changes in key personnel in customer's organization

5. Recent or projected significant changes in contract
   a. Greatly increased or diminished scope of mission
   b. Significant changes in level of expertise required
   c. Consolidation or segmentation of contract

6. Political factors (at high level)

7. Cost factors
   a. Analysis of incumbent contract costs versus your company's estimated cost
   b. Incumbent G&A structure
   c. Competitors' history on cost competition
   d. Contract profit potential. Is it *worth* winning?
   e. Estimated cost of preparing proposal

8. Long-range benefits to your company (other than monetary)

9. Can suitable subcontractors or teaming arrangements be obtained if required?

10. Do you have the resources of time, money, and talent to prepare a winning proposal in the allotted time frame?

Incumbent's Performance Ratings

Several years ago, Congress at the behest of the news media's "investigative reporters," who were seeking easier ways to satisfy the "public's right to know," passed the Freedom of Information Act. Little did the Congress or the news media realize the consequence of their act! For in the ensuing years, it turned out that 80 to 90 percent of the requests submitted to government officials for information came from business—big business, little business, and medium-size business—all supplicating for information on government contracts. Now you can demand a copy of any government contract (subject to security classifications and, to a limited extent, company proprietary information). A smart marketeer can usually deduce such information as the number of people on the contract, the cost to the government of each man-hour of labor performed, the award fee contracts, which are becoming the most common), the contractor's performance ratings. Now, for example, if you know the incumbent contractor's performance ratings have been 97 percent, 98 percent, and so forth, for the past two years and you cannot underbid him on cost by at least 10 percent, forget it. Don't waste your bidding and proposal money, because you haven't got a chance. If his performance ratings started out bad five years ago but have steadily advanced to, say, 90 percent, then proceed with caution. Let me hasten to add, however, that performance ratings are only one of many factors to consider in making the Bid/No Bid decision. Sometimes the ratings themselves are deliberately skewed (politics), sometimes an ambitious contracting officer will award to the lowest bidder, no matter what. So consider other factors.

Incumbent's Program Manager

The program manager is, of course, the most visible personality in the performance of the contract. Sometimes a dynamic and highly personable program manager can develop such close personal ties with the customer that he can strongly influence the retention of a contract. Sometimes he can make the contract performance look better than it really is. Sometimes his influence may be limited to just one, or two, of the most influential of the customer's top people. In that case you must focus on these people, or, as an alternative, pray for their early retirement.

One very important thing to look for is the history of program management. Did the incumbent actually provide the manager whose résumé was submitted in the proposal? Has the same manager been on the program for the past two years or more? If the current manager has been there only a year or so, what became of his predecessor—kicked upstairs to a nothing job? or sent packing to an assignment in Timbuktu? or sent out to pasture in ignominious defeat? Of course, these are clues as to how successful management has been in establishing good customer relations. I once insisted on bidding on a contract that no one in my management wanted to bid on, because the incumbent had been there "forever." All other potential bidders felt the same way. But I knew that the incumbent had gone through three program managers in the past two years. For that reason alone, I just *knew* the government was ready to change contractors. (There had to be something wrong.) And I was right! We won the contract. There were no other bidders (except the incumbent), of course.

Procurement Office History on Recompetitions

Prudent decision makers should begin with the reliable premise that there is a vast amount of difference between procurement officers, whether they be governmental or commercial. That is to say that fundamental differences in operating philosophies and procurement practices do prevail, depending on where you are doing business.

Let us take the Department of Defense (DOD) Agencies, for instance. With most Navy procurements, the proposals usually resolve themselves into a résumé competition. He who produces the résumés that most closely fit the job descriptions or job requirements set out in the RFP is the most technically qualified. I have encountered few Navy RFPs that emphasize management techniques or technical approach in their criteria for a winning proposal. The Army, on the other hand, emphasizes technical approach—how you are going to perform the technical aspects of the contract. The Air Force similarly emphasizes these criteria, but also places more stress on management techniques—reporting procedures, job control, maintenance procedures and documentation, quality assurance, and in short, the "paper work" aspects of the contract. NASA, in my opinion, is the most demanding and meticulous of all, stressing all of the above in a very orderly and systematic manner. These are, of course, generalized statements, subject to exceptions and are to be interpreted in that context. Also, these are my personal observations, based on some 20 years of working in the field—not on the periphery as a manager, but down in the trenches writing the proposals.

So far we have been talking about generalized differences between different government agencies. But there is also a marked difference in the methodology employed by different procurement offices within the same government department, or between one commercial firm and another. Some will play everything strictly by the book, others will play favorites, others will play games—pitting one contractor against another. Others will pick the low-bid contractor even when the low bid may be ridiculous. (Never mind that this sort of thing may signal fraud, collusion, naiveté, or ignorance, or all of the above to an honest government investigation. It happens.) The point is that you, the decision maker, must analyze and be aware, *and beware* of such situations, lest you be the innocent (and poorer but wiser) dupe in these little games that bureaucrats play (and business executives too).

This awareness (some people call it intuition) is what makes good business managers good, and this is what good marketeers are for—to find out these little idiosyncrasies about the customer before your company foolishly commits bundles of money to a lost cause.

Changes in Key Personnel in Customer's Organization

In the earlier section on Incumbent's Program Manager, I alluded to the fact that a personable program manager can strongly influence a contract award by developing close personal ties with certain influential customer personnel. I know of one case where a $100-million contract was competed, and *no one* except the incumbent bid on it. The next time it was competed, no less than ten contractors took a shot at it. The reason is not hard to figure out. During the period between the two competitions, all of the key personnel on the customer's side had left the scene, and the program manager of the incumbent had moved on as well. Nothing else had changed. The scope of the contract was virtually the same—the type and location of the work, the number of people involved, the potential for profit, the competence of the incumbent in performance of the contract—all had remained virtually unchanged. What else do you suppose accounted for this sudden interest in going after this contract? The astute decision maker takes a good hard look at these changes in key personnel, and sometimes this element becomes the single overriding factor in making his or her decision.

Sometimes you have the situation in a services contract where the incumbent has been on the contract so long that an "incestuous" relationship develops. After 20 years or so, the contractor's kids grow up and marry the customer's kids and, of course, get jobs on the contract. The customer's personnel retire one day and they go to work for the contractor the next day . . . and so on. Here you have a situation where it would almost take a sudden mass exodus of customer personnel to get a change of contractors, because there are so many of the customer's decision-making personnel who have a stake in keeping the incumbent contractor. What you must look for then are political changes in this particular environment. If this is a government contract, have there been any charges of favoritism in the news media? Have any local congressmen been identified with suspiciously close ties to the key civil servants involved? Has anyone been asking questions in Congress about the remarkable longevity of this incumbent contractor? Have there been recent significant changes at the department head level in Washington? If it is a commercial contract, has there been any recent change in management philosophy (for example, a sudden need for more economical operations by demanding *real* competitive bidding by their contractors)?

One must not give up just because the incumbent contractor has been there "forever." Sooner or later, one or more of the factors I am outlining here will catch up with them. It happens all the time. I know of some cases where even the local congressmen could not save the day for the incumbent. A new administration, a powerful new department head forced a change, because the incumbent got arrogant (had been there so long he thought the customer was working for him), complacent and smug (slow to respond to customer requirements), lazy, careless, or whatever. Along comes a new decision maker and it's a whole new ball game.

Recent or Projected Significant
Changes in the Contract

Frequently, a significant change in the scope of the contract will provide a vital clue to the customer's intentions or become a vital factor in making the Bid/No Bid decision.

Let's say, for openers, you are looking at a government contract that has been won by open competition. Now the government is changing it to a Small Business set-aside. If you are not a small business, well, forget it. If you are, go for it—especially if the incumbent is big business. Obviously there is going to be a change of contractors.

Let's say you are looking at a contract that involves about 30 man-years per year. But you know there are some big programs coming down the pike that can run this effort up to 700 man-years. Does the incumbent have the resources to absorb this much responsibility—the personnel, the management, the risk capital, the experience? Does he have enough smart people around to write a winning proposal of this magnitude? (It takes a lot more smart people to write a 700-man-year proposal than a 70-man-year proposal.) If the answers to these questions are No, then you can probably clobber him. If they are No, but the customer likes him, then you should approach him with a teaming arrangement or a prime/subcontractor proposition. You can both make it easier on yourselves that way.

Let's say you are looking at the opposite situation—a contract that involves 700 people but is expected to decline to about 70 people. First, you have to decide if it is worth bothering with financially, but we will discuss that later. One must ask: What are the reasons for the diminished scope? A temporary lack of funds to proceed with the program? A transfer of the responsibility, in large part, to another agency or to another part of the economy? A hiatus between programs? Temporary political considerations? A reorganization into a higher technology operation? A good executive must look down the road to three or five years hence, if he is to make the right decision. That declining little old 70-man contract could bounce back to become a gold mine. And, here is another consideration to think about. Those 700 people you inherit—you might pick up some great talent there already trained for jobs you have been trying to fill elsewhere in your company. Have you accumulated data on what it costs to recruit engineers and good technicians lately? Think about it.

Now let's consider another factor regarding significant changes in the contract—the level of expertise required. I know of a case where the incumbent had been firmly in the saddle for years, running a fairly routine operation. There he sat—fat, dumb, and happy, knowing he had it all wrapped up for years to come. But revolutionary developments were in the offing. A very high technology requirement was coming down the road for the design, installation, check-out, and operation of a sophisticated command and control system for support of the space shuttle program. Could the customer trust the incumbent to shift into high gear to support this kind of transition? Apparently not, because out of nowhere came a high technology outfit that stole the contract from under the incumbent's very nose, to the surprise and consternation of the whole aerospace community. This was another example of few companies bidding, because they thought the incumbent had it made. They failed to consider the vital factor that a radical change in the expertise required may signal the demise of the incumbent contractor.

When you see a situation like this you must start asking some questions. Like, does the incumbent have the technical resources to make the transition? What is his reputation in the technical community? Especially, what kind of management does he have? Dynamic, flexible, resourceful, and imaginative? Or stodgy, slothful, rigid, and incompetent? If the latter, go for it, and full speed ahead, damn the torpedoes!

Another important factor to be considered regarding changes to the contract is consolidation of functions formerly assigned to other contractors, and the reverse: segmentation of the contract into smaller bites, thus breaking it up into more contractors. This is NASA's favorite game. They oscillate in almost predictable periodic fashion from one extreme to the other. I would not presume to fathom the NASA mentality, but I suspect they save a lot of money this way, because it does engender more competition—honest competition—between contractors.

This is how it works. Contractor A has been the incumbent for eight years. Contractor B (and maybe C and D) come to NASA (and/or their Congressmen) and whine, "We would like to bid on this contract, but it has grown so big and complex in eight years that it is hard for *any company* to bid on it." And NASA says "Not to worry, we are going to break this contract up into 'more manageable segments,' to encourage freer competition." They break up Contractor A's piece of the pie into three parts (some of which may be a small business set-aside—this is good politics), and all the suckers rush to *buy in* to get a piece of the action. Then five or ten years later, regardless of how good the incumbent's performance has been, NASA decides that the three contracts (plus some more, perhaps) should now be combined into one big contract. Here come all the suckers again to buy in to this big contract, figuring to break even after a couple of years and start making some money off it in about five to six years, which is about when NASA suddenly decides that it is better, after all, to segment this contract into 'more manageable parts' again. As you can see, this does promote competition and probably saves the government money as long as contractors continue to fall for this con game. Never mind that the poor guys who work in the trenches, the engineers and technicians on the NASA program, end up working for 20 different contractors in their careers. And, until the Service Contract Act was passed, every time a new contractor came in these poor guys took a pay cut to enable the new contractor to "buy in" without losing money, and corporate executives got a promotion for winning new business.

Anyhow, this book is not concerned with what should be, but what is, and the lesson here is: Whenever you see a consolidation of contracts or the reverse, a segmentation of a contract, you can be sure there is a business opportunity there. There will be new contractor(s) there. You can bet on it.

Political Factors at Higher Levels

I touched on this subject before when discussing the course of action when an incestuous relationship has developed between the contractor and a customer. This subject deserves more attention, however. Political relationships can be subtle traps for the unwary. I once knew of a company that hired a recently retired Air Force general as a corporate vice president. Good politics, huh? Only trouble is, this company, which had been very nearly dependent on Navy contracts, did not realize that this general had mercilessly slashed Navy R&D budgets while he served on a budget committee with DOD. Gradually this company's Navy contracts began magically to disappear.

More than once I have seen companies waste a lot of money bidding on a contract to be awarded by a government agency that was at the time being sued in court by the company—or the reverse—being sued by the government agency that was their potential customer. Don't kid yourself! It doesn't work. Bureaucrats are human, too, and they have long memories, especially for their enemies (forget the due-process stuff). Axiom: Don't sue a government agency or even protest a contract award unless you are prepared to stop doing business with that agency again. They will close ranks against you like the bulls in a herd of cattle when a wolf appears. And it will all be perfectly legal, too.

If your company has a history of labor problems, beware of bidding in a labor sensitive area. No project manager (customer) wants to see his operation shut down by a strike.

If you are in the habit of making large political contributions or aligning yourself with one side or the other, or with the truth were known, politics is probably the most important factor in the decline and fall of the Jim Ling empire (Ling-Temco-Vought). He contributed heavily to the Democratic Party campaign fund just before Richard Nixon was elected president. Maybe it was just coincidence that the government filed an antitrust suit against him shortly after Nixon took office. My advice to business executives is to stay away from politics. If you want that kind of high stakes gambling, go to Las Vegas. All you lose there is your money.

Cost Factors

Analysis of Incumbent Contractors
Versus Your Company's Estimated Costs

You remember, I said a good marketeer, using the Freedom of Information Act and other sources, could find out just about everything you want to know about the incumbent's contract costs? Well, if your marketeers *cannot* find out what is the competition's labor costs, general and administrative costs (G&A), fee, cost per man-hours, and so forth, then *fire them all* and get some new ones with some smarts. Any good marketeer should be able to get you all this information plus the actual number of people employed on the contract. Now figure up what it will cost your company to perform this contract. Total up your G&A, fee (that you can live with), and unit labor cost times number of people that realistically can perform the contract. If you find you are way out of the ballpark, forget it. (Some companies can and will perform more cheaply than others.) If you are almost in the ballpark, then consider the various alternatives for underbidding the incumbent, for example, effecting realistic economies in staffing without degrading performance. But remember, your proposal must justify—present a credible rationale—for effecting these economies without degrading performance. If you are now in the ballpark, then consider some other cost factors.

Incumbent G&A Structure

Various contractors play all kinds of games with G&A. Is the incumbent's G&A realistic or some kind of hocus pocus financial sleight of hand? Maybe you can use some of this hocus pocus yourself. I mean, for example, having three G&A pools: one for bidding on government contracts, one for commercial contracts, and one for operating the company overhead structure.

Incumbent History on Cost Competition

Some companies are very rigid, for example, persuade the customer to stipulate ceilings on G&A—which amounts to *normalizing* G&A. "Normalizing" means, in effect: Make everybody bid the same G&A. Some companies will settle for zero fee or even a negative fee (you owe the customer if your performance is rated below a certain percent, say 60 percent). Some are going to make what they consider a fair profit—"or take your contract and stick it. . . ." Of course, some will simply bid at a loss—a losing contract is better than no contract at all, I guess. I know one company that even found a way to lose money on a cost plus fixed fee contract!

You must make an informed assessment of your competition—not just the incumbent but all the rest. Are you prepared to play this hand? Is it worth it?

Contract Profit Potential

Is it worth winning? First, just a few words on the three different types of contracts variously employed:

1. *Cost Plus Fixed Fee (CPFF).* You bid a certain cost, the customer negotiates the actual cost and then you get a fixed previously agreed-upon fee. (Most companies usually spend this money on new furniture for the executive suite, or if more progressive, on bidding other contracts.)

2. *Cost Plus Award Fee (CPAF).* You live within a negotiated cost (that is, you are supposed to), then you get a variable fee, based on the customer's evaluation of your performance. (Of course, this evaluation is inherently subjective, so oftentimes there is a lot of hanky-panky here.)

3. *Firm Fixed Price.* Obviously this is the type that the customer likes best, because it exposes him to the least risk and you to the most. Since you are taking all the risk, if you are smart, you bid higher than you would on a cost plus contract. Who can foretell all the disastrous unforeseen contingencies that might ensue before performance is completed? Strikes, material shortages, accidents, law suits, and so forth, and so on. But many contractors rush in, happy and carefree, to bid on these traps, and the bleached bones of their bankrupt companies strew the landscape from coast to coast.

Estimated Cost of Preparing Proposal

Most people, new to this business, are appalled at the high cost of preparing a *good* proposal. (Of course, you can save a lot of money by doing a bad proposal.) One wonders how you can go through about $200K just writing a 300- to 400-page document. Here are some of the things you have to consider:

- Travel and per diem (for people you have to bring in to help write the proposal)
- Overtime (for nonexempt employees)
- Printing and publishing costs
- Consultant costs
- Travel and per diem for bidders conference, bidders tour, Q&A session, best and final offer, and negotiations
- Marketing costs—the up-front people who make the customer contacts and gather data on the contract. (These costs can run very high if the customer is one of those who likes to be entertained.)
- Advertising—Nothing demoralizes an incumbent like seeing a full-page ad by a competitor looking for all the skills he is currently utilizing on the contract.
- Lease costs—Usually you have to lease separate facilities for writing the proposal for a variety of reasons, security for one. The security aspect cannot be overemphasized. You must treat your proposal material and the discussions

involved therein about the same as you would TOP SECRET material. I know of cases where someone paid janitors to dump all the office trash from a competitor's wastebaskets into his station wagon. You can find out a lot of information this way if you don't mind sifting through the coffee grounds, and the like, and if you like to piece torn scraps of paper together. Of course, that is a mean, dastardly trick, isn't it? But it's a cruel world out there, and it happens. So much for security.

- Office space—You must also be very circumspect about using your leased office space (paid for by the customer) for writing your proposal. This is most definitely frowned upon. It's not worth jeopardizing a contract in the hand for another one that is still in the bush, so my advice is: Don't do it. Go out and lease some office space elsewhere.

This should give you some idea of the costs involved and why it runs so high. You can take this as a guide: I usually start with a bid and proposal (B&P) cost of about 1 percent of the gross value of the contract for one year. For example, if you have a $100 million contract for five years, one year would be $20 million, and 1 percent of that would be $200,000.

I once worked for an outfit that was pretty big, mainly because they had a huge number of small contracts. Finally, they decided to go for it—a big contract worth $80 million (five years). I told them it should cost about $104K, a conservative estimate, but a realistic one if they had proceeded in an orderly and systematic manner. Well, they thought my estimate was ridiculously high. The *vice president* took me to lunch and eyeballed me indulgently, "We don't do things that way here, Don." "I know you don't, you've never bid on a contract this big before." "Well, we don't have $104K to write a proposal." "Then you better forget it. If you can't stand the heat, stay out of the kitchen." They went ahead and bid on it, and given the naiveté displayed above, you can readily see that it was *not* in an orderly and systematic manner. That is the kind of situation—confusion followed by panic—that runs your proposal costs up dramatically. The last I heard, they had spent over $196K . . . and still climbing.

I hope I have made my point. Writing a proposal is expensive, even if done efficiently. If it is not done efficiently, you can easily double the cost. Later on in this book, we will get into the various ways of doing it efficiently. I can say it all here in just one phrase, "Get organized and get prepared *before* the RFP comes out."

Long-Range Benefits to Your Company
(Other than Money)

Of course, there are often some contracts that are worth risking a loss for the first year or two, but you musk ask yourself, *Will the contract give me an entry into a new geographic area? Is there a realistic possibility for growth?* Many business opportunities are just deadends. You start out with a 50-man-year effort and you know that five years from now it will still be a 50-man-year effort. You may make some money out of it, but growthwise there is no long-range benefit to your company. So, of course you don't bid them on a break-even or low-fee basis— unless you are really desperate. Like the company president's Cadillac is beginning to show some wear and tear.

There are a number of business opportunities around these days as a result of OMB circular A-76, which required government agencies to contract out services currently performed by the civil service or uniformed military personnel. Most of those are of the ash and trash variety (mundane, labor intensive, non-technical), but a few bear investigation by even the technically oriented (engineering) firms. Sometimes the government decides to make the transition from a government- to contractor-operated facility, a gradual process, allowing the normal attrition of civil service personnel (retirement, transfers) to be filled by contractor personnel. I wrote a proposal once for a contract calling for the ridiculous sum of 14 man-years of effort. But we knew what the government had in mind and before long this 14 had grown to over 50 and eventually will be some 200 or 300 people performing all the technical support for a large National Range.

Would it broaden your business base?

I once knew a company that had gradually accumulated a variety of small contracts over the years. Those involved specialized technical support for engineering development, followed, in some cases, by O&M, of communications and sensor systems—such things as microwave, telemetry, radar, CCTV, command and control systems, display systems, and the like. If you put all these things together (plus photo-instrumentation) you have all the ingredients of any missile range, bombing and gunnery ranges, and military test centers. After some astute marketing people on their staff convinced top management of this obvious situation, they decided to go for it by bidding an O&M contract on a whole range, not just the bits and pieces as they had been doing. With a little expert consultant help, they were able to write a credible proposal and win this contract. This company whose top management didn't know how to spell range, has an entry into a whole new world—the National Range System, which numbers 13 National Ranges plus some 30 test centers operated by the DOD. Here is a prime example of how, with a little luck and willingness to take a chance, you can tremendously broaden your business base in one fell swoop.

Would it open the doors to a new technology?

Would it get you on the ground floor of a rapidly expanding technology or commercial application? Let us say, for example, you have been doing some work in bioresearch, maybe a contract with the Environmental Protection Agency and perhaps a grant from a medical foundation. You have a small but brilliant staff of biochemists, biophysicists, and medical research people. Some of your work has brought you to the periphery of genetic engineering. Along comes an opportunity to jump into this fantastic field with both feet. Go for it—even if you have to mortgage the office furniture to get in, even if the work is being conducted in the Mojave Desert, even if you have to subcontract to someone else, even if you don't expect to make any money for ten years. You are getting into an esoteric field with commercial possibilities, the limits of which are literally inconceivable. Genetic engineering will probably change our planet in the next 50 years. Of course, you are not going to turn a fast buck here, but think of it this way: you are paying for an education that will eventually repay you a hundredfold.

Would it enhance the reputation of your company?

Let's say you are a company primarily in the business of performing engineering services. You have a number of engineers and software development people with degrees working for you. You are enjoying slow but steady growth in a medium- to high-technology area. You see an opportunity to win a large contract to operate a large warehousing and storage facility for the government—a typical example of what is scornfully termed in the aerospace community an "ash and trash" job. Of course, it's not your bag, but you feel confident you could win it and it will help pay the rent. Don't do it. Just a few contracts like this and you will soon be known in your technical community as an ash and trash contractor, and you will not be taken seriously by the customers.

On the other hand, say you are looking at a business opportunity in your field—right down your alley. It's a bigger job to tackle than you are accustomed to doing, and it will cost you more to write a good proposal than you really care to spend. Never mind, go for it! If you win it, you will have established your reputation in the community. Even if you lose, and you write an impressive proposal, you will be remembered by the customer next time around—with respect.

Will the contract give you an entry into a new geographical area?

Suppose you have a number of contracts on the West Coast. A big opportunity is about to arise in Hawaii. You know that Hawaii is 2000 miles away; you don't have a base from which to operate there. It will therefore cost you more money to invade this market area. On the other hand, the opportunity is in a technical area in which you specialize; there are many other potential opportunities in the same field that will arise in the future. If you win this contract, you will have a base from which to expand in the Hawaiian and mid-Pacific areas. Do you play it safe (and maybe stagnate), or do you take the gamble?

Would it eliminate competitors from the area?

This could be just the reverse situation. You already have most of the business in this technical area in Hawaii—except for one upstart outfit based on the West Coast whose one contract in Hawaii is coming up for bid soon. You don't really want the contract. The customer is notoriously penurious, constantly complaining and nitpicking your performance, and a real pain in the neck for any manager. You will probably have to rotate managers every six months and give them combat pay! You don't believe there are such customers? Then you haven't been around much. Well, what do you do? You grit your teeth and bite the bullet, because if you win it, you will have complete domination of this geographical area, and competitors will all have the disadvantage of having to develop contacts and establish a base from 2000 miles away.

Can suitable subcontractors or teaming arrangements be obtained if required?

First, let's take a quick look at the whole concept of using subcontractors. As a general rule, I would say, "If you can show credible experience in handling 80 percent of the work involved, don't subcontract." Get consultants to help you on the other 20 percent of the technical proposal, if necessary, but go for the whole ball of wax. If you have expertise in only 60 percent or less, then you had better look for a subcontractor or teaming arrangement or forget the whole thing. If it's between 60 percent and 80 percent, you have a decision to make taking into account all the other factors involved on a case-by-case basis.

The tendency of most amateur managers is to subcontract everything in sight that diverges in the slightest with what they are currently doing. I once worked with an outfit that was hung up on subcontracting. The RFP was like a shopping list for subcontractors: "Here is a requirement for road maintenance. We'd better get Pan Am or Bechtel to help out on that." (The "road maintenance" turned out to be a situation where you went out and hired someone with a road grader two to three times a year to smooth out the road.) "Here is O&M of a radar; we'd better check with RCA." (The radar was an old beat-up FPS-16 that at least 10,000 retired master sergeants or Navy chiefs could operate in their sleep.) "Here is a TM station; we have never done O&M *on the front end* of a TM station; maybe we should No Bid." (This company had designed and installed, but not *maintained*, TM systems all over the country.) I told them, "Nobody gives a damn about front end-rear end, just show them with a few paragraphs in the technical proposal you know how to do the mundane work of O&M on a TM station. Hire someone in the area to grade the roads when needed. Hire some E-techs with radar experience, and so on, and so on." Yet some of these idiots were still talking Pan Am, RCA, and more, two weeks after the RFP came out.

Generally speaking, if the customer *wanted* two (or more) contractors, he would put two RFPs out on the street as separate contracts. A good and practical reason for this reluctance to accept subcontractors is that there is no privity of contract between a customer and his contractor's subcontractor. He must deal directly only with the prime contractor; he has no legal standing whatsoever with the sub. If he is dissatisfied with the performance of the sub, he cannot get rid of him. He has to intimidate the prime into getting rid of the sub or else terminate the prime contract. Obviously, one can get into all kinds of legal entaglements here. A smart customer wants to keep things simple—fixed responsibility in one person, the program manager. He doesn't want to hear any excuses about how the *sub* screwed up. He wants an integrated team responsible directly through one individual that he can put the collar on if things begin to go wrong.

There is one broad exception to this general rule, and this is the result of the well-intentioned legislation conferring special advantages (such as "set asides") to minority businesses, small businesses, labor surplus area businesses, and the like, in order to achieve greater participation by such businesses in the American dream. But, as often happens with good intentions, the reality is somewhat different from the intentions. In other words, if you want the contract you *must* find some small or minority-owned business to subcontract part of the work to. It doesn't matter if *none of the employees* of this minority firm are members of a minority, just so the business itself is minority owned. Of couse, it's good business to team up with these small minority businesses, because many government contracts are set aside exclusively for bidding by them. If you want the work, you have to get them to *subcontract it to you.* Never mind that the minority business may consist of only two people working out of a hotel room. They own the business and sub-contract all the work to you, so everyone is happy—except perhaps the few taxpayers who know what is going on.

Do you have the resources of time, money, and talent to prepare a winning proposal in the allotted time frame?

Time, money, talent. These are the only resources you need to write a winning proposal! It is interesting to note that the amount of money required is inversely proportional to the amount of time and talent available:

$$\frac{\text{Proposal}}{\text{Expense}} \propto \frac{1}{\text{Time} + \text{Talent}}$$

I'll explain the derivation of this theorem in a later chapter. What we are concerned with here in making a Bid/No Bid decision is whether or not, under the existing circumstances at the time, these resources are available to an adequate extent or not?

A well-managed company should establish long-range marketing plans based upon knowledge of scheduled future business opportunities. Then, after applying the Bid/No Bid criteria outlined in this chapter, you must prioritize these opportunities in such a way that the necessary resources will be available at the proper time. Sometimes, two great opportunities will hit the street at the same time. You must make a choice between one or the other. The overzealous or naive manager often tries to go for both at once, hopelessly diluting his resources and failing miserably on *both* opportunities.

Maybe you have just won a big contract and you have assigned all of your best technical talent as key personnel on the new contract. Now another big opportunity comes along. Well, you don't start putting these people to work on another proposal. They've got their hands full getting the new contract started off on the right foot. Don't jeopardize the bird in the hand for one in the bush. Don't get greedy. Pass this one by, or else demand *and get* substantial, qualified, competent support from corporate headquarters. And remember—this is going to cost you.

There are situations where you have ample time and money but inadequate talent. You might have some great engineers that really know the job, but none of them knows how to write. Some of the best engineers I've seen could not write a postcard home to Momma without getting it all fouled up.

You might have plenty of time, money, and even a few articulate engineers but no one around who knows how to write or manage a proposal. The talents required to prepare a winning proposal come in many forms: organizers, managers, engineering/scientific experts; people with ability to write fluently, logically, and persuasively; a supporting cast of characters: typists, word processors, editors, graphic-arts people, reproduction people, administrative-control people, security specialists, costing and pricing specialists, marketing strategists, marketing-intelligence people, image makers, *and* top management that has the wisdom to either be helpful or stay out of everybody's way.

As you may conclude from the many preceding pages devoted to the Bid/No Bid criteria, this is a major factor in the proposal cycle, not to be taken lightly. I hope the point has been made, however, that the Bid/No Bid decision, like all

DAVID A. WATTERS

important business decisions, can be reduced to a systematic thought process enabling managers to make a sound decision arrived at by research, weighing of significant factors, balancing of risks; and then there is no looking back.

PLANNING AND GETTING ORGANIZED

The proposal effort begins with planning and organization. This should be accomplished primarily by the proposal manager, who, incidentally, should ideally be the proposed program manager of the target contract. He or she should develop a proposal plan that outlines the entire proposal effort, based upon information available at that time. That would include organization of the proposal team, tentative selection of key personnel on the contract, basic working assumptions, and *modus operandi*, that is, where and how the work will be done, tentative schedules, security measures, typing, editing, reproduction procedures—in other words, a complete plan based upon such information as is currently available. (When the RFP is released, this plan is then updated, based on the RFP, including a general outline of the proposal.) The proposal manager should also gather together all the reference material available in anticipation of the next phase, which is the research and preparation phase.

How do you organize a proposal team? First, it is imperative that management provide all the resources and support that you, the proposal manager, need to do your job and then get out of your way and let you manage the proposal. This subject will be covered in detail in Chapter 6. The important thing to understand at this point is that the proposal manager must decentralize control into manageable segments. That is, you should have a deputy to provide the detailed guidance and review required for any major proposal. You should have an administrative coordinator to take care of the myriad of administrative details required in putting a proposal together. Then you must have team leaders to guide and supervise the technical details of writing the various segments of the proposal.

A good way to approach this is to analyze the job to be performed—the contract—and then make a list of all key personnel it will require to perform the contract. For example:

> Program manager
> Operations manager
> Engineering manager
> Maintenance manager
> Quality control specialist
> Administrator

Now you have pretty much listed who your team leaders are going to be. One will be responsible for everything in the RFP concerning operations, one for engineering, one for the management plan (the administrator), and so forth. These are your leaders.

I worked on a proposal recently where there were 18 key personnel. That is too many team leaders. You can't have that many people reporting to you, so you combine functions so that you have no more than ten people at the most reporting to you. If some of the team leaders want to allocate some of their functions to subteam leaders, that is their business. Don't try to tell them how to do what you have assigned them to do.

Another way you can organize the proposal team is to divide the RFP into certain areas of responsibility and assign each to a team leader. But the trouble with that is that you want to get the proposal team organized *before* the RFP comes out. This is so that the team leaders can begin the extensive research and analysis without which any major proposal effort is doomed to failure.

RESEARCH AND PREPARATION PHASE

The next stop, the *research and preparation phase*, is an opportunity to bring all proposal contributors to the point where they can project themselves into the customer's environment. This might be compared to the work of a good intelligence officer preparing his unit to operate in the enemy's environment. A good G-2 will find out everything there is to know about the enemy—all the enemy's strengths, weaknesses, leaders, methods of operation, objectives, plans, aspirations, and decision-making apparatus. Once he has learned all these things he must disseminate all this knowledge to the troops in time to be useful for the next phase of operations.

If, for example, this is a data center O&M job, the proposal team should be capable of vicariously experiencing a day of working in this particular data center by the time they are through researching this program. This can be done by studying any existing documents such as the old RFP, descriptive material such as PR brochures, articles in trade and professional journals, old proposals, background material provided by Marketing, and so forth; also briefings by consultants or ex-employees on this contract (to include any slides or vue-graphs they may have, showing facilities, equipment, and the like), analysis of the existing contract including all modifications, impressions of visitors and/or users of the facility, and so forth. When all this material has been absorbed, the proposal team should get together and compare notes, discuss courses of action, tentative organization, tentative approaches; in other words, brainstorm it to death! If this has been accomplished exhaustively and conscientiously, it should be very easy to project oneself into the target environment. *Only then* can you begin to write with authority.

Far in advance of the RFP release date, drafts of certain parts of the proposal should be prepared. You *know* that the RFP is going to require a section on company experience, résumés of key personnel, and company administrative policies. Put these things together now. The format may have to be changed, but you will have all the information assembled when the RFP comes out. Then you can finish up these items early and put them to bed, so that you can concentrate on the more important things during the limited time available after RFP release. Also, if you have properly done your homework, as described here in the research and preparation phase, you can write the first draft of the executive summary at this time.

And, finally (during the last ten days before RFP release, say), the proposal team should have a short course in proposal writing, ideally by your corporate in-house proposal expert. Every company should have at least one person who is an expert on proposal preparation. Even if he has to be flown out from corporate headquarters, the expense is well worthwhile if he can at least get everyone started off on the right foot and pointed in the right direction instead of wandering around, as is usually the case.

What I mean by "wandering around" is illustrated by a classic case of mismanagement, confusion, disorganization, and ultimate panic that I once had the misfortune of experiencing on a proposal effort. First, there was no effort whatsoever to organize a proposal team or even select a proposal manager until a week *after* the RFP came out! This was on an eight-week turn-around for a $70-million contract! Even though I had gathered a vast amount of research material on this particular contract, there was no one around to study or analyze it, because no one had been made responsible for the proposal, and of course, no proposal team had been formed. When the RFP came out, the top management went into action. A cast of thousands suddenly converged from all corners of the United States to become instant experts on preparing this proposal at a cost that would have made Cecil B. DeMille blanch. But, here they came, leaderless, confused, unbriefed, some sullen and resentful, others happily looking forward to getting away from home and living it up in a fun place in expensive hotels. A couple of people were sent to the bidders conference and bidders tour, but whatever they learned was forever locked in their psyches, because neither one ever gave a debriefing or a written report on the information presented by the customer. (It turned out that even though they had been given a tape recorder to tape the proceedings, no one had ever bothered to transcribe the tapes for the poor, beleaguered "herd" of proposal contributors.) I guess you have the picture. The proposal ended in the predictable panic situation. Just for graphic examples: the cover letter contained a gross error in common grammar. One of the tab dividers that was supposed to be entitled "ENGRG SVCS" came out "ENGRG SUCS"—and so on.

Now, the message is: This is what happens when you fail to make the proper preparations and fail to get organized *before* the RFP comes out. This sort of nonsense will happen every time, no matter what the situation or the people. Given the same conditions, it is as predictable as $Ag^+ + HCl \rightarrow AgCl \downarrow$, and this is why I must emphasize that the preliminary preparations I set forth in this book—research, organization, analysis, and proposal preparation training—are vital to successful proposal writing. And, by using the information this book provides, you can ensure the achievement of these objectives.

DEVELOPMENT OF THEMES

An essential part of your homework should be the formulation of themes that will be woven into the fabric of your proposal. Themes are the basis for establishing your competitors' weaknesses and your company's competence to perform the contract. The preliminary preparation for the composing of any good proposal should include the formulation of these themes, which of course are based primarily on good G-2 work by your company's marketeers. This is to say that you must find out what are your competitors' weaknesses (especially the incumbent since he has an advantage) and then try to offset these weaknesses by demonstrating your company's strong points in these areas by good technical approaches and company-related experience. This is definitely an important factor in preparing for this proposal, and there will be more on this subject in Chapter 6, where the proposal manager's responsibilities in this area are described.

On the other hand, one must not get too bogged down in the pursuit of themes. You must concentrate on adequately responding to the RFP. This is your primary objective. Your proposal is going to be evaluated by professionals. Amateurs and charlatans usually pitch the high-pressure sales pitch aspect of proposal preparation as all-important. It is not, and you will find out why in Chapter 5.

Now, back to the subject at hand. What are some examples of themes? If you really do some digging, you can always find some areas of performance where the incumbent is weak. If, after conscientious digging, you can't, then you should consider No Bidding the contract.

Here are some examples of themes I developed on a recent proposal involving engineering support of a government-operated facility.

Quick response to changing requirements. Quick response to engineering development requirements was hampered by the incumbent's lack of degreed engineers who could prepare profesionally competent engineering studies to get an engineering development project off the ground. Mostly, the "engineers" were good nuts-and-bolts supertechs.

Control, monitoring, and status reporting of work in progress. There were no formal professional-level procedures for controlling work progress, cost control, project status, or MIS systems to maintain and report status of projects.

Property accountability and inventory control. The incumbent used antiquated methods and manual procedures for this function, resulting in slower response time and excessive man-hours of labor.

Smugness, complacency, even arrogance, by incumbent personnel. (The longer the same contractor retains incumbency the more this situation is likely to develop.)

It doesn't take a genius to see how you can take advantage of all these incumbent deficiencies by emphasizing and taking special care in addressing these areas in your own proposal. But, you must do it with subtlety and discretion. Take the positive approach: emphasize *your own* capabilities in these areas, not the incumbent's failures. (That could kill you.) Prove to them how you are going to provide quick response capability, efficient and reliable status reporting, and inventory control.

Project eagerness and enthusiasm—with a touch of obsequiousness thrown in for good measure.

For example, take the first theme mentioned, "quick response," and so forth. Throughout the proposal, you need to emphasize the great engineering talent your company has, how you would assign degreed engineers from your vast resources of talent to perform every appropriate task. You have a wealth of experience in performing engineering studies and R&D studies, and in successfully managing engineering development projects similar to this particular one. Cite examples of successfully completed projects that were finished within the allotted time frame.

Don't try to address a theme all in one place. You must sprinkle the theme statements throughout the proposal, so that all the proposal evaluators get the message. Work your themes into the various sections of the proposal where they fit, but don't get strident about it. *Keep it subtle.* If you lay it on too thick, I guarantee it will be counterproductive.

CHAPTER TWO

PLAN OF ATTACK FOR RESPONDING TO REQUEST FOR PROPOSAL

This chapter is presented essentially to aid the poor, tormented engineers, mathematicians, and scientists to get started off properly, systematically, and without confusion in the alien world of proposal writing. These are the people who are down in the trenches of the proposal war, who provide the real guts of the proposal. While their work may not seem glamorous, it is definitely challenging to do it right. And they may all take a fierce pride in knowing that without their expertise, their unique talents, and their conscientious devotion to duty, no proposals would ever be written, no new business acquired; their company would stagnate, and all the vice presidents, general managers, marketing specialists, CEOs—the whole lot of them would be out on the street eventually, looking for a job. For in the last analysis, these engineer-level people are the only ones who understand the technical complexities and ramifications of the job well enough to write authoritatively about it.

Let us review for a moment the general phases involved in proposal preparation:

First: The research and decision phase presented in Chapter 1 (pre-RFP activity).

Second: RFP release. At this point, the proposal team has had an opportunity to study all pertinent documents related to this project, and now the RFP is out. *You are ready to go.*

Third: Preparation of the proposal. But wait a minute! Do they expect you to read all of that stuff? It must be two feet thick. It would take a strong man just to carry it into the next room. And look, it seems to be written in bureaucratese—a language you never learned in college. It seems *hopelessly* disorganized; you don't know where to start; you need a plan of attack.

I use the word "attack" advisedly. Many people, when called upon to contribute to a proposal effort are intimidated when they see that voluminous RFP.

I say, "Take a positive attitude. Attack! Don't wait for it to attack you." Proposal writing need not be an ordeal. If the proper preliminary preparations have been made, and if your efforts are properly organized and everyone proceeds in a systematic and orderly manner, you will probably be surprised at how easy it is.

The key to success is:

1. Adequate preparation before the RFP is out.
2. Adequate preparation before starting to write.
3. Intelligent constructive review and guidance as you go.

ELEMENTS OF A CONTRACT

Before we go any farther, perhaps we should take a look at just what is a contract, since our ultimate objective here is the obtaining of a contract. Basically, a contract requires three elements: an offer, an acceptance, and a consideration (money or a promise to do or *not* to do something). A request for proposal (RFP) is neither an offer nor an acceptance. It is a request by the customer for you to make an offer. If the customer accepts your offer (your proposal), you have a contract, provided there is full agreement on the consideration. Oftentimes, the customer accepts your offer by telling you that your proposal is most acceptable of all those tendered and then offers to negotiate the price. You may go to an enormous amount of work writing a proposal, but remember, this is just an offer by you; the customer is not obligated to accept it and under common law rules, does not even have to give you a reason *why*. For Government contracts, however, the Defense Acquisition Regulations (DARs) and other government regulations do require the government to give you a debriefing on demand, if the potential contract is for more than a stipulated amount of money (depending on the type of contract).

Those are the very basic elements involved. Of course, there are all sorts of things that can complicate the whole process and even make a completed contract null and void—things like fraud, collusion, misrepresentations, not negotiating in good faith, and so on—all of which is beyond the scope of this book. I bring the subject up here because of the confusion that exists among many people regarding the status of the RFP. Strictly speaking, the RFP is not a contractual document; it is simply a request by the customer for someone to make an offer. Your offer (proposal) is an integral part of any contract that may result, and that is why precision of language is of paramount importance in writing proposals. Anything you say you will do in the proposal constitutes a promise to perform specifically as stated, and the offeree (customer) can take you to court and nail you for damages in the event of your failure to perform as stated. By the way, the only excuses accepted are those attributable to an act of God. If you can't perform, you had better pray for a tornado or hurricane, or at least, a well-placed bolt of lightning.

Another facet of contract law that is worthy of note here is the verbal contract. Of course, I'm not talking about the contract resulting from an RFP. I am talking about modifications thereto that have the same legal status as a contract. Generally speaking, verbal contracts are not enforceable. There can be exceptions to this rule when a court, applying rules of equity, will, under certain circumstances, enforce a verbal agreement where one side of the agreement has in good faith been fully and unilaterally performed. That is why contracting officers get very nervous about subordinates or field managers telling the contractor what to do or how to do it, or in any way even implying that the contractor is authorized to perform something different from what he was contracted to do. Most contracts have a clause specifically forbidding the contractor from departing from the terms of the contract without specific *written* authorization from the contracting officer. Even that does not necessarily guarantee against problems arising from misunderstandings, misinterpretations, and excessive "supervision" by the customer. The point I am making again is that those misunderstandings (so-called "gray areas") and confusion can best be avoided by careful, precise language in writing the proposal.

One last item: personal service contracts. The specter of the personal services contract casts its long shadow over the entire government services field, striking terror into the hearts of bureaucrats who would foolishly arrogate themselves into the function of managing people instead of managing contractors. The DAR, and other government regulations, severely limit the authority of the government to negotiate personal services—that is, limited to a few narrowly defined exceptions to the general rule. The general rule, of course, is that personal services contracts are illegal. This rule, wisely supported by the courts and Congress, is intended to keep bureaucrats out of the business of supervising the contractor's employees, but instead, force them to deal only with contractor management according to well-established contract management procedures. One of the basic reasons for this is that, otherwise, bureaucrats could circumvent civil service ceilings by augmenting their numbers with contractor personnel, turning the whole system of government service contracting into a sham. Also, the basic purpose of government is to govern, not to perform services that can be performed more reliably and cost effectively by contractors who have learned how to survive in the competitive world.

NASA, in particular, has had a difficult time with this rule. In their zeal for empire building, they like to overlay the entire contractor's organization with a skeleton organization of their own—with monitors down to the Tech B level. Naturally, this kind of micromanagement has gotten them into trouble and into court. One of the results of this was the establishment of the famous Pellerzi Criteria, Pellerzi being the judge who tried to set up some guidelines for government managers to help them stay out of "personal services" trouble. Since this is not a book on industrial relations, I won't list these criteria here, but you could ask any NASA industrial relation's specialist. I'm sure they are all well-read on the subject.

HOW TO USE THE RFP TO IMPROVE YOUR PROPOSAL

The first thing you notice about most RFPs is their bulk. Don't let it intimidate you. If it is a government procurement, most of it is pure eyewash, and you don't even have to read it. Probably the main thing that gives it the most bulk is the recital of government regulations that are meant to restrain all of us greedy, unscrupulous, racist, union-busting, macho, unpatriotic contractors. For instance, we have the certificate of non-segregated facilities, the Buy American Act, EEO requirements, SBA requirements, Davis-Bacon Act, Baker-Nunn Act, Affirmative Action Compliance, labor surplus area status, clean air and water certification, and so on and so on, *ad nauseam*, and now even a "female-owned" business regulation! The best thing about this part of the RFP is that you don't have to read it. The proposal manager, and especially the cost proposal manager, however, do have to leaf through all this morass of bureaucratic obfuscation in order to ensure that all the required certificates are rendered (plus the various other items that pop up from time to time throughout the RFP) and that all requirements are addressed.

What you *should* read are: the technical description, if any; the proposal instructions (with annexes); the evaluation criteria; the statement of work or specifications, and the schedule. Do not bother to wade through the rest unless your proposal manager directs your attention to other specific items. Figure 2-1 shows you which parts of the RFP you must read.

All RFPs have similar formats. (See Figure 2-1, p. 40.)

There are many ways you can use the RFP to improve your chances of winning. First, there are always annexes and/or appendices in any good RFP that provide an insight into the nature of the hardware, facilities, software, and *modus operandi* of the operation. You must study these carefully. First, give them a cursory going over to find out what information is there, then really analyze the portions that pertain to your section. You will also usually find such things as workload data, clues as to staffing, work breakdown structure, support services available, and so forth. It is absolutely imperative that you make use of this data and weave it into your proposal. To ignore it means certain disaster.

Second, you must make note of the format the customer uses to present the RFP—the sequence in which each of the technical requirements is described, the implied relative importance of the various segments. Even the punctuation, use of abbreviations, paragraph numbering, and the like, are

FIGURE 2-1 Example of RFP Format

	Part I.	General Instructions
		Section A: Cover Sheet, Technical Description
		Section B: Boiler Plate
		Section C: Detailed Instructions, Conditions, and Notices to Offerors
Normally these are the only items the proposal contributor need be concerned with.		Annexes: Proposal Instructions
		Scope of Work (or Statement of Work)
		Contract Data Requirements List (CDRL)
		Data Item Description
		Section D: Evaluation Criteria
	Part II.	The Schedule (Line Items, Model Contract)
		Section E: Supplies, Services, and Prices
		Section F: Inspection and Acceptance
		Section G: Deliveries and Performance
		Section H: Special Provisions (For example, Definitions, Options, GFE, Patents, and so forth)
		Section I: Contract Administration Data
		Section J: Packaging and Marking
		Section K: General Provisions
		Section L: List of Documents
	Part III:	Appendices

important because you must write a proposal that as closely as possible mirrors the RFP. One of the reasons you must do this is to make it as easy as possible for the evaluators to find what they are looking for (and we will be getting into more on this later).

Usually, the RFP will list the evaluation criteria that will apply to your proposal. The customer also generates detailed "standards" based upon these evaluation criteria, but you will never see these in an RFP; they are for the use of the evaluators only. (This whole process is described in detail in Chapter 5.) It is vital to the success of your proposal that these evaluation criteria be foremost in your mind as you write. You must ensure that each and every one of these criteria is addressed in your proposal with the greatest emphasis given to the most heavily weighted criteria. It is also imperative that you make a logical deduction as to what the customer will generate as *standards* upon which to evaluate the presentation you have made to fulfill these criteria.

Next, check out the contract data requirements list (CDRL) or sometimes just contract data list (CDL). Many people have a tendency to slough over this section as just so much more boiler plate. *Don't do it.* You should not only address each respective CDRL item in your write-up, but also, where appropriate, show how this report is generated and acknowledge that it will be done. After all, remember, your offer (proposal) becomes part of a real live contract and if you do not acknowledge responsibility for doing it, you are not contracting to do it. *And*, the customer can therefore evaluate your proposal as nonresponsive.

Look over the Data Item Descriptions. These are put in the RFP to help you, not to be ignored. (Sometimes they are not included in the RFP and you will have to do some research to get them. *Do it.*) You must know the data item descriptions, so that you can adequately describe what you have to do and how to do it, to acquire and report the data.

As you can readily see, the first step in preparing to respond to an RFP involves a substantial amount of study and analysis. You must do this in a systematic manner *before you start to write a single word.* The novice proposal writer will grab a pencil every time and start writing a concoction of gibberish, repeating sentences out of the RFP, reciting how accomplished his company is, how they did a similar job in Timbuktu, assuring the reader he knows how vital this program is to the customer's needs and so on . . . zzz. Read on and learn how to avoid making a fool of yourself.

THE MOST IMPORTANT STEP IN PROPOSAL WRITING

Analyzing the RFP

What you are doing now is the most important step in proposal preparation, because if you fail to analyze the RFP properly, you are bound to fail to respond to it—no matter how well you do everything else. If you fail to respond in any particular, you will have a deficiency. Too many deficiencies and your proposal winds up in the wastebasket and your company will soon find you a nice job in Angola, or maybe on the north slope in Alaska—one of those offers you can't refuse.

Below is a check list for analyzing an RFP, followed by *specific instructions* for working your way through this check list.

Steps in Analyzing the RFP

You do not have to read *all* of the RFP.

Scan through it all the way, page by page. Mark any pertinent paragraphs.

Then concentrate on the following—read these sections carefully:

Technical description
Statement of work or scope of work (SOW)
Instructions of offerors
The schedule (contract line items)
The model contract (if any)
Contract data requirements lists (CRDL)
Evaluation criteria

Then *study* and *dissect* your section of SOW.

Take one sentence at a time.
Look for hidden meanings.
What are the ramifications?
Exactly what do they mean?
Exactly what do they want?
Exactly what do they need?
Mentally formulate your response.
A lot of brainstorming with colleagues here will pay dividends.

Now compare your tentative response to:
  A. Evaluation criteria. Try to deduce Government "standards." (See Chapter 5 on government standards.)
  B. Proposal instructions. (Have you touched all the bases?)

First—read carefully through all the pertinent sections: technical description, SOW, proposal instructions, the schedule, the model contract, and the evaluation criteria. Give special attention to the SOW instructions and evaluation criteria, for these are the driving forces that control your proposal. Mark the paragraphs that pertain directly to your area of responsibility.

Take one sentence at a time. Keep asking yourself, "What do they really mean by that? Could there be any other meanings? *Exactly what* does 'span of control' mean in this case? *Exactly what* all is comprised in 'system integration' here?" Two heads are better than one. You *must* get a competent colleague to help you interpret. You would be surprised at how many interpretations can be extracted from a seemingly innocuous sentence. Have some brainstorming sessions with others on the proposal team. Then sketch out with a few notes, the salient features of your response. *Do not try to write anything yet!*

Now bounce your tentative response against the proposal instructions and the evaluation criteria. Have you left anything out? One way to do this is to write the heading of each of the evaluation criteria in turn, and then underneath this heading write key words that indicate your approach to fulfill each of these criteria. For example, refer to Figure 2-2 (Excerpt from a DOD RFP). Suppose you are now writing the "Microwave O & M" part of "communications." How will you address "perception of problem areas, etc." to show your expertise in this area? List your ideas for accomplishing that. Then, how will you show the evaluator you know how to supervise this segment of the program? And so on through the criteria.

Now you are ready to make a detailed outline of your response. Remember, your detailed outline must conform to the proposal general outline that should be provided to you by your proposal manager a few days after RFP release. The general outline will usually be pretty much dictated by the proposal instructions. The detailed outline will require con-

FIGURE 2-2 Excerpt from a DOD RFP

CRITERIA FOR TECHNICAL EVALUATION OF QUOTE IN ORDER OF IMPORTANCE (RFP)

1. Engineering and analysis
2. Management and support
3. Instrumentation operation and maintenance
4. Communications

Within Each Functional Area

1. Approach
   Perception of problem areas and sound solutions
   Management/supervision
   Feasiblity of technical approach to each task
   Extent of support resources (vehicles, housing, etc., required)
   Degree of assurance that the bidder can actually perform
   Promotion of efficiency in accomplishment of tasks
2. Personnel
   Quantities and skills proposed; qualifications
3. Experience
   Developmental engineering; procurement of complex equipment

---

siderably more thought and in-depth analysis. A good way to start, however, is to extract the pertinent part of the proposal instructions and the SOW (that is, the part that is pertinent to the proposal segment you have been assigned) and carefully go through it, underlining all the key *nouns* together with their modifiers. You will usually find that these nouns, taken in sequence, practically dictate your outline for you.

For example, study the excerpt from instructions from a recent NASA RFP. The key words and phrases are underlined. See Figure 2-3 (Excerpt from NASA RFP). You will note that paragraph (2)(a) is concerned with "Organization and Policies" (general heading). But here are the key words:

mission contract requirements	(line 2)
integrated management system	(line 4)
requirements of a mission contract	(line 5)

FIGURE 2-3 Excerpt from NASA RFP

---

(2) Organization and Policies

    (a) The proposer shall describe his organization and management policies to accomplish the <u>mission contract requirements.</u> A thorough treatment of these items shall be provided to enable an understanding of the proposer's rationale for an <u>integrated management system</u> specifically tailored to the requirements of <u>a mission contract.</u>

    (b) The proposer shall provide <u>organization charts</u> which show the entire proposed organizational structure, including relationship to the <u>corporate and/or division organization.</u> Complete rationale for the structure can be provided. A description of the proposer's <u>internal lines of responsibility and authority,</u> and the <u>interface relationships</u> with the Government and any subcontractors shall be shown. The <u>interrelationship</u> of the planning, operational, and management responsibilities of the various <u>organizational elements,</u> including element size, shall be provided. <u>Organizational features</u> which contribute to maintaining flexibility and efficiency throughout the performance of the contract requirements shall be described. If work is performed by the contractor for commercial organizations or other Government agencies, the extent to which this proposed effort will be <u>organizationally integrated</u> with that work, and supporting rationale, shall be provided.

    (c) The proposer shall provide <u>functional policies, techniques, and procedures</u> applicable to the management of the mission contract effort. Such policies as <u>delegation of authority and/or responsibility,</u> <u>degree of management autonomy,</u> and subcontract management, which the proposer would require to adequately direct the overall contract effort, shall be included. The proposer is encouraged to show evidence of <u>proven management policies and procedures</u> that are effectively operating within his parent organization or on other contract effort that can be applied to the particular demand of this contract.

Now you have to conclude that in paragraph (2)(a) they want to see what you know about managing a *mission* contract. Tell them what you know about organization and management policies as pertains to managing not "a" but *"this"* mission contract. Don't tell them *anything else* in this paragraph. Now, take paragraph (2)(b):

organization charts	(line 1)
organizational structure	(line 2)
corporate and/or division organization	(line 3)
internal lines of responsibility and authority	(line 5)
interface relationships	(line 5)
interrelationship of the . . .	(line 7)
organizational elements	(line 8)
organizational features	(line 9)
organizationally integrated	(line 14)

Obviously you had better talk about organization and interfaces in this paragraph and *nothing else.* Furthermore, you must address each of these items insofar as logically possible in the same order that they appear. In other words, for example, don't talk about interfaces before you talk about organizational structure.

Now take paragraph (2)(c):

functional policies, techniques, procedures	(line 1)
delegation of authority and/or responsibility	(line 3)
degree of management autonomy	(line 4)
subcontract management	(line 4)
proven management policies and procedures	(line 7)

As you can readily see, they want you to talk about management policies, techniques, and procedures here, not *particularly* as pertains to mission contract—you covered that in the first paragraph; not as regards organization—you covered that in the second paragraph.

The main subheadings under paragraph (2) might look like this:

(2) (a) Mission organization
    (b) Organizational structure
    (c) Management policies and procedures

Of course, I've left out the subheadings under (a), (b), and (c), but you can readily see how they would be constructed from the listings above. Each subparagraph would consist of one or a combination of these listed items, preferably in the same order as they appear in the RFP.

HOW TO EVALUATE YOUR OUTLINE

First, check the sequence of your outline. Have you presented each item strictly in the same sequence presented in the RFP (specifically the SOW)? Reread the proposal instructions. Have you conformed to these instructions thoroughly and to the letter? Now reread the SOW and other technical informational material. Have you left anything out? It's best to make a check list of all the items required by the RFP pertaining to your section, and then check them off against your outline. This check list must include:

CDRL items, data item descriptions

Applicable model contract items

SOW items

Other items that may apply to your area that are sometimes found elsewhere in the RFP

Evaluation criteria. Does your outline ensure that all evaluation criteria will be addressed?

When you have completed this check list, find some knowledgeable colleagues or colleague and go over it with them. They may have some suggestions. Discuss the approaches you are taking and the technical decisions you propose to incorporate in your write-up. Discuss with them any alternative approaches you may have considered and rejected. If doubts or disagreements arise at this time, fall back and regroup. Check the proposal instructions again, the technical description, the scope of work. Perhaps more research is indicated; maybe you should consult with outside sources and technical literature that has been written on the subject. Don't be in too much of a hurry to get on with writing the proposal. What you are doing here is the very foundation of the proposal and you can't expect the final product to be acceptable if it rests on a faulty foundation. If you get this part of the work done perfectly, the rest will be easy. *Don't rush it.*

If, after all, you still have some doubts or disagreements, confusion or whatever, get together with your team leader or proposal manager (as the case may be), and try to resolve the difficulty. The time to do it is *now*. He may not have time to help you later. Furthermore, the difficulty you may be having may be shared by others. Possibly everyone may have the same problem, but no one realizes it yet. Possibly the difficulty you are having could affect the quality of the entire proposal, so the proposal manager should know about it as early in the game as possible. Possibly, it could arise from a flaw in the RFP, and the customer should be apprised of it at once, so that he can issue a modification.

After you have resolved all problems, gone over your check lists, discussed with colleagues to corroborate the feasibility of your approach, and otherwise assured yourself that your outline is perfect—the very best that you can do—then, and only then, turn it in to your proposal manager for his approval.

Summary: Preparations for Writing the Proposal

1. Re-read any RFP informational material (technical description) or outside references bearing on this section.
2. Check for any applicable CDRL items.
3. Check for any applicable model contract items.
4. Check data item descriptions.
   a. Have these been incorporated into your outline?
5. Does the outline address the items/factors in the same order that they appear in the RFP?
6. Go over your outline with one or two knowledgeable colleagues. Discuss approaches you have taken and technical decisions/alternatives involved in your approach.
7. If disagreements or doubts arise over any approach taken, study the RFP requirement and/or other references again.

Sometimes the correct solution may be inferred from the language of the RFP. If there are still doubts or disagreements, discuss them with the proposal manager and get them resolved. *Now*!!!

8. Find a quiet place to work where you can project yourself into the customer's environment and write without interruption or distraction. This is important, because proposal teams often are assigned one big room where there are a dozen conversations going on at once and 16 telephones ringing, and before you know it the whole thing degenerates into one big bull session. Nobody can accomplish serious writing or analytical thinking under those conditions.

The foregoing simply lists a few reminders to ensure that your outline is complete and in conformance with the RFP. If you have any uncertainties at this point on decisions you have made, interpretations of the RFP, or approaches you have taken, by all means, get together with your proposal manager and get them resolved at once. Don't try to do anything else until you are satisfied that this outline is perfect, and it has been approved by the proposal manager.

WHAT IS THE FLAVOR OF YOUR PROPOSAL INPUTS?

Most people would say vanilla. I would say a little more than that, maybe chocolate chip. But before I get all tangled up in my own metaphor here, let me hasten to say that it's a different kind of flavor I'm talking about. I'm talking about the flavor the dictionary describes as, "a particular quality noticeable in something, as in a style of writing, painting . . ."

It is premature at this point to get into specific details, such as usage and writing styles (that subject will be covered in Chapter 4). You have completed your outline, however, had it approved by the proposal manager and are, at last, now ready to start writing. I'm sure you are just dying to know what do you do next, and how do you get started. First of all, believe it or not, if you have done your work perfectly up to this point, the hardest and most significant part of your effort is already behind you. All you have to do now is fill in the blanks.

In very general terms, for now: Your style should be informative without being pedantic; factual without being boring; logically organized without being rigid; thorough to the extent of showing the readers everything they want to see, and no more. Thought processes should flow—generally, from the generic to the specific; from the known to the unknown; from the simple to the complex; and from the past, to the present, to the future. The whole purpose of the proposal is to convince the reader that you know and understand the problem and know what to do about it. You would be amazed at how many people in this business do not understand this. Some people think you can overwhelm the reader with eloquence, or recitation of past accomplishments, or other tangential discourses. First and foremost, you are *responding* to an RFP, and you must do so with an economy of words. Many technical people have a tendency to assume the reader knows all about the subject, so they don't want to bore him with the details. They forget the purpose of the proposal is to *let him know that you know* these details—not educate him.

I recently worked on a proposal that included the operation and maintenance of a TM (telemetry) station. They dragooned a super tech (that's an engineer without a degree) to write the input for the proposal. He came to my office sullen, querulous—the typical "Why me?" attitude. (Like most engineers and technicians, he hated to write.) I gently explained what we needed. (I say "gently," because he was all we had, and besides, he was indeed an expert in this field.) He grudgingly took his extract of the SOW and proposal instructions and grumbled his way out the door. A few days later I got the usual mishmash one gets from someone who had had no training or experience in proposal writing: some cut-and-paste stuff obviously from an O&M manual, some regurgitation of the RFP, and a description of a TM station we once operated in Virginia. I read it over and called him and asked in a nice way to come back and let's talk.

I didn't say anything about his unsuitable input, but instead got him talking about the way this proposed TM station was equipped. It was obvious that he had much enthusiasm for telemetry operations and was really expert and knowledgeable. At that point I put on a dumb act (which was easy for me), and asked, "I've always wondered just what they mean by 'stripping the data.' Could you explain that to me?" His eyes lit up and he went to the chalkboard and gave me a 20-minute dissertation on how TM equipment works, how the system is checked out, data collected, stripped, and so forth. When he got through I said, "Now *that* is what I need to see in your input for this proposal." "But, they already know all that," he said. "Maybe so," I said, "but the idea is, the customer wants us to prove to him that *we know that*!" "OK, no sweat," and he disappeared. A few days later I got one of the best inputs of the whole proposal.

This is because his input was informative, factual, and to the point. He wrote with authority, because he was an expert in his field. He didn't waste words rambling or beating around the bush, or trying to "snow" the customer with verbosity, because he knew exactly what he wanted to say, that is, to tell the customer how this telemetry facility works and how to operate and maintain it. When you really know your subject, you can tell about it with few words; the subject matter is already organized in your mind. Your writing will have the proper flavor. When you don't really know your subject, you will have a tendency to ramble, to throw in extraneous and irrelevant material, and in short, try to "snow" the reader. I assure you the proposal evaluators are going to recognize this fraud every time. You are winging it and they know it. The bottom line is that you have to research your subject thoroughly and organize it carefully *before* you start to write.

CHAPTER THREE

HOW PROPOSALS ARE ORGANIZED

Nearly every proposal should have four (or more) main parts: executive summary, management proposal, technical proposal, and cost proposal.

INTRODUCTION

Why do I emphasize repeatedly that you must *not* start writing without going through all the preliminary steps to ensure that you are thoroughly prepared? Because of something called "pride of authorship." It seems that once someone has committed an idea on paper, he feels a sense of having created something—something that is a part of his very self—his ego. No matter how half-baked, inept, useless, and irrelevant it is, he develops a fierce, defensive, emotional involvement with it. To destroy it is like tearing a babe from its mother's arms. The less occasion the proposal manager has to do this, the less trauma one is going to suffer in the course of writing the proposal and the less heartburn for the proposal manager. We want you to be so thoroughly prepared before you start writing that your second corrected draft can be considered a final draft (ready for management review).

Here is what you must bear in mind as the ultimate objective of what you are doing.

Technical Proposal Objective

Your objective as contributor to the proposal is to give the people on the evaluation board a warm comfortable feeling that "here at last is a company that understands our problems and knows exactly what to do about them."

Commit this to memory. This says it all. Of course it doesn't say *how* to achieve this worthy end. That's what you are going to learn in the remainder of this book.

The general outline for the proposal, which must be provided to the proposal team by the proposal manager soon after the request for proposal (RFP) is released, must conform to the format directed or inferred from the RFP. What you as a proposal writer need to know is that your detailed outline must, in turn, reflect the format directed or inferred from the RFP. Your detailed outline should, insofar as possible, follow the same sequence, even within each paragraph, as it appears in the RFP. Remember, the whole idea is to make it as easy as possible for the evaluators to find what they are looking for in your proposal.

The proposal instructions (often entitled something like "Instructions to Offerors") will usually set out clear instructions regarding the format desired, including sequence of items to be addressed. If not, you must infer the desired format from the sequence that the items are presented in the RFP and the format (spacing, organization, terminology, and so forth) used in the RFP.

Some proposal managers or company executives get the idea that there is only one way to write a proposal—their way. For example, "Every proposal should start out with the management plan and end with a phase-in plan." And so they decree this format in every proposal no matter what the RFP instructions say. Nonsense! If the customer's staff wanted this sequence, they would have constructed the RFP in that sequence. Remember, you are responding to an RFP, not indulging in an exercise in creative writing. Responding to the RFP in the same sequence as it is presented makes it easy for the evaluators to find what they are looking for. Why make their job more difficult?

Do not think you know better than the customer, what he wants. You must follow the instructions to the letter (or if no instructions, follow the RFP format, no matter how unorthodox it may seem). For example, if they address company experience, technical approach, résumés, and management plan, in that order, give them a proposal in that order. How you have done it before, or how *you* think it should be organized, is irrelevant.

As with every rule, this one must be tempered with common sense. As Emerson said (among other things), "Foolish consistency is the hobgoblin of little minds. . . ." Sometimes the customer, too, makes mistakes. Say, for example, they mention QC as one of the responsibilities of the contractor and then, two paragraphs later, having forgotten they already listed QC, they bring it up again. Don't repeat your write-up on QC. And *never* split a single subject into two different places. The only occasion when you would talk about QC twice, for instance, is if you are talking about two different kinds of QC—where there is a clear distinction between the two, so that they are in reality two different subjects.

It has become common practice to include an executive summary at the beginning of a proposal, whether the RFP requests one or not. (Of course, if they request one and want it at the end of the proposal, OK. You give them an executive summary at the end of the proposal.)

The reason it has become common practice to include an executive summary is that this is the only effective means of delivering a message to the top brass involved in the contract award decision. Obviously these people are not going to read the whole proposal. They must depend on the recommendations of the team of experts they have selected to analyze the proposals. The executive summary is a means of bypassing these experts and getting to the top decision maker and influencing him in making the contract award decision.

Most proposals will essentially be comprised of four sections in two separate volumes. The first volume includes the executive summary, the technical proposal and the management proposal. The second volume is the cost proposal.

The cost proposal is nearly always separate from the rest of the proposal. The reason for this is twofold: (1) The necessity of treating cost data with greater confidentiality than technical data (except, of course, where technical data is classified); and, of equal importance, (2) the need to maintain objectivity by the evaluators of the proposal. Engineers must not be biased by cost data in drawing conclusions regarding engineering judgments, and accountants must not get involved in making engineering judgments.

The technical proposal describes your methodology, your approach, your concept for accomplishing the purpose of the contract. It may or may not be the most important part of the proposal. You have to check the evaluation criteria that are usually published in the RFP. Sometimes the management proposal will be the most heavily weighted, depending on the special circumstances surrounding this particular contract. Sometimes, for example, if the customer has been having a bad time with management problems and no trouble with operations problems in the past, he will weight the management area higher than the technical. In most cases, however, the technical proposal will be the most heavily weighted—and for good reason. This is where you have to prove to the customer you know how to do the job. I know of one case, in fact, where the technical proposal was the only section that was even evaluated (because the customer was in a hurry to make an award). I think the customer in this particular case has lived to regret this decision, because the contractor he picked didn't know how to manage, as he was soon to prove by his performance.

The management proposal customarily includes a *pot pourri* of obligatory loose ends: résumés, job qualifications, supply procedures, payroll and accounting, safety, security, quality assurance, personnel administration, and just about anything else that doesn't fit elsewhere. Many of these items are virtually "boiler plate"; that is, they can be mostly cut and pasted from other proposals or company SOPs. The wise proposal manager would assign someone to start putting all of this stuff together before the RFP comes out, so he can put it to bed early and devote his time to the important parts of the proposal. Of course, one of the most important, if not the *most* important part of the proposal, is where you tell the customer *how* you are going to manage the contract. The way many proposal managers go wrong is in thinking you can handle that part the same way you handled the miscellaneous loose-ends part, that is, cutting and pasting, generalities, vapid clichés, and so forth. The customer wants to know in specific terms, exactly how you are going to manage *this* contract—generalities, he does not need.

As stated in the introduction, the cost proposal is beyond the scope of this book. Suffice it to say that costing is almost always done by experts in that field (as it should be), but there must be close coordination between the technical and cost people (they are commonly referred to affectionately as "bean counters"). The costing people must rely completely on the technical people for staffing and technical support requirements of the contract. It has been my observation that generally there is a woeful lack of communication between these two essential elements involved in a successful proposal.

Summary: Organization of Proposals

1. Nearly all proposals will be organized into four sections, usually in two volumes:
   a. Volume I
      Section 1. Executive summary
      Section 2. Management proposal
      Section 3. Technical proposal
   b. Volume II
      Section 4. Cost proposal
2. The management proposal will usually include company experience, administration (personnel, supply, accounting), safety, recruiting, quality assurance, résumés, and so forth.
3. The technical proposal—how you do the job.
4. RFP instructions usually direct (or imply) the format to be followed.
5. Do *not* think you know what the customer wants better than he does.

EXECUTIVE SUMMARY: WHAT IT IS, WHO WRITES IT, AND WHEN

The executive summary is the first section of your proposal, *always*. It should be prepared by the proposal manager in close collaboration with the cognizant individual from marketing. It is the one place in the proposal where you have a chance to make an out-and-out sales pitch, so make it good. I don't mean to encourage any shrill, high-pressure harangues, or any clever, and slick Madison Avenue gimmickry, either. You are not selling cars or two-pants suits. You are talking to seasoned, somewhat cynical professionals. There is no way you are going to overwhelm them except with facts and impeccable logic. So—Rule No. 1 here . . . *Don't get cute!*

The sales pitch should set the theme—the message that is woven through the entire proposal. Here is an example of a message. You know the incumbent contractor has been floundering on some of his engineering projects because he often fails to have the necessary materials and equipment available on site when they are needed. So you stress the fact that *you* have an automated supply/resupply system, a job control system that tracks the progress of all projects in progress, and that every important project is PERT (Program Evaluation and Review Techniques) charted, so that the project manager knows exactly what is the status of any facet of a project at any time. You emphasize this capability and cite real-world examples of it in your executive summary. (More on this in the section on proposal management.) That is why the executive summary should be written early in the proposal cycle, so that the various contributors can be guided by its themes and weave this message into their own contributions.

The executive summary should be dynamic, terse, and designed to capture and hold the interest of the nontechnical reader. *Do not get bogged down in detail.* No busy charts, graphs, complicated algorithms, tables, and so forth. Keep it simple, direct, and to the point. Use words that have impact. Keep sentences short. Just one central thought to each paragraph. The entire executive summary should usually be no more than six pages—never more than ten. For most proposals, if you use more than ten pages, the chances are you really don't know what you are trying to say or how to say it, or worse, your summary will have lost its focus.

A word about the writing techniques for executive summaries. Get everything up front. This is your opportunity to tell the top brass about the super quality of your company's performance, about the devilishly clever, innovative solutions you have developed to make the customer's life happy and carefree. Just knowing you will be on the job, solving all the technical problems expeditiously and reliably will give the customer a warm feeling of confidence. Don't hold back any goodies for some other section. Don't try to tantalize him with hints of what is to come or try to keep him in suspense. You are not writing a novel. This is not a treasure hunt. You have one chance here to influence the most important minds engaged in the selection process: the chairman of the Evaluation Board, the contracting officer, the company president, or (for Government procurements) the source selection authority. This is the one part of the proposal these individuals are likely to read. *Don't blow it!*

A good executive summary should be a kind of condensed version of the proposal with a sales pitch included. Most people I've seen who try to write executive summaries get carried away with the sales pitch part and don't ever get around to telling the customer anything else. It doesn't take much empathizing here to realize that too much sales pitch becomes counterproductive—turns the customer off. At a certain point, sales pitch turns into bravado and *braggadocio*, and instead of impressing the customer, you end up making him hate you (personally). You especially must avoid bragging about dollar profits, dollar growth in profit, or anything else to do with profit. The customer would prefer to read that you are a *non*-profit organization. The inescapable conclusion to be drawn from data showing high profit growth is that you are somehow ripping off your customers. In the government, many civil service people regard *all* profit as immoral and suspect. Most bureaucrats actually believe that *any* profit from a government contract is a rip-off.

We've talked about what not to put in the executive summary; what *do* we put in it? First, you want to show that you understand this contract and know how to do the work. You identify the critical areas involved in successful overall performance of the contract. This is where you really put those "themes" developed by your marketing people to good use. You concentrate on the known vulnerabilities of the incumbent and the known salient features of your company. You know the incumbent has made mistakes somewhere (everybody makes mistakes); you know he has weaknesses somewhere. (If you don't know, you haven't done your homework; time to move back to square one and go through the Bid/No Bid criteria again.) Having addressed the subject area where you know the incumbent is weak or made mistakes, you rub a little salt in the wound by pointing out how important this area is to success of the contract, and then you *prove* in as few words as possible that *you* know how to handle this problem. I say "prove," because using bare, unsupported statements just won't do it. The reader may not believe you. If space simply does not permit proof (the executive summary *must* be brief), then, at least, reference the part of the proposal where you do prove you can handle the problem.

Another thing that any good executive summary should have is a brief outline of the unique, innovative features of your approach to managing or performing this contract. What? You don't have any innovative features or new ideas to present? Back to square one. You haven't done your homework. You can't expect to win the contract by just proposing to do the same things the incumbent is doing. Why would the customer want to change contractors?

I recently worked on a proposal involving a sizable number of people in a remote overseas location supporting missile launch and re-entry operations with a sophisticated array of advanced electronic and electro-optical instrumentation. The workload, that is, support of missile operations, had temporarily, but indefinitely, dwindled down to the point where most of these people were sitting around twiddling their thumbs most of the time. What you have here is an overseas, intermittent operation, requiring continuous maintenance of sophisticated electronics and electro-optical equipment, punctuated by periods of intensive operational requirements. When the contract came up for recompetition, most bidders simply proposed the same outmoded *modus operandi* the incumbent was using, with somewhat reduced staffing through trimming some of the fat here and there. But this is a situation begging for innovative ideas. How about an intensive cross-training and cross-utilization program so that all maintenance people become operators during mission support? How about subcontracting some of the intermittent work to be performed on an *ad hoc* basis by a small company based in the United States? How about exploring the feasibility of drastically reducing the on-site force to a skeleton maintenance crew and flying people out to augment this force on a TDY basis to support specific operations? How about furloughing some of the people (forced leave of absence) on half pay when the site was scheduled to be inactive for over a month? These are the kinds of innovative ideas the successful contractors must come up with in order to win. Every proposal calls for some brainstorming and imaginative thinking. No idea should be dismissed simply because "they have never done it that way." No one should be put down for suggesting something that seems far out. You must have these brainstorming sessions throughout the proposal cycle, but the best time to start is well before the RFP comes out, using the information you have at hand and, if possible, extrapolating from there. This is the way you bring problems to the surface, questions that require answers, problems you never thought of before, uncertainties, and imponderables. If you have these sessions *before* the RFP release you still have time to do some research, pulse your sources of information, and stimulate your marketing people to get answers. (That is what they are there for—make them earn their pay.) If you wait until the RFP comes out, you won't have time to do all these things.

So much for innovative ideas in your technical or management approach. Just remember, you must get these major ideas right up front in your technical or management approach. They must be ideas of major significance. Don't bore the reader with trivialities. You *must* keep his attention with dynamic style and attention-getting ideas.

Why are you best qualified to perform this contract? If you have done your job well and kept the reader's attention up to this point, this is the question he is now asking. This is the time—not sooner, not later—to give him the all-out sales pitch. Now you can brag unabashedly about your great conquests and your fantastic achievements of the past and present. *But keep it brief.* The details are presented in another part of the proposal. The executive summary is like a résumé. You give them just enough to whet their appetite, pique their interest, make them curious to know more. Don't show the warts, and most of all, don't bore them with details, or you will lose them.

Weave your themes into your description of innovative ideas. Remember themes should be woven into the entire proposal, not segregated in just one place. You subtly emphasize the areas where you can provide superior performance to that of the incumbent, or you relate your past performance or expertise in accomplishing similar work where you know the customer has the greatest anxiety. It's important that you do this in a positive manner. Sure, you know that the incumbent's engineering projects never seem to get off the ground, but you don't talk about these failures. Talk about your demonstrated successes in these areas, and briefly how you have managed to cope with these problems. You know that one of the other bidders has been accused of cheating the customer, that charges have appeared in the newspapers, that court action is threatened. Never, never mention these things openly. Just point out how your company has devised built-in safeguards and feedback systems to prevent this sort of thing from happening. Let the reader make his own connection to what has appeared in the news media.

Perhaps you have a dandy solution to a problem in management or technical design that neither the customer nor other contractors have thought of before. Trouble is, it would cost the customer an outlay of money to implement this solution. Never mind, you don't have to cost this outlay into your proposal. Just give him a peek at what you are prepared to do *after* you get the contract. For an investment in your idea the customer can realize an ultimate cost savings of many times the investment or get a system design that will solve problems he had always thought were impossible. Again remember, you must do this briefly. As I said, just give a peek. The executive summary is not the place to come up with a systems design. Just give the reader the parameters of what makes your company the best qualified to do the job.

How is your proposal organized? A brief answer to this question is usually the last part of the executive summary. This is where you provide just a few brief paragraphs to walk the reader through the proposal—the sequence of your presentation, any ground rules; for example, "Additional résumés will be included in the Appendix." Point out any special efforts you have made to help the evaluators find things—organization charts printed on tab dividers, complete staffing data provided as a fold-out at the end of each volume, anything of general interest that may be helpful in evaluating your proposal.

Now you have completed the executive summary. As I said before, it should add up to about six to ten pages. If you use more than ten, the chances are you haven't thought through what you have to say.

One last observation before we leave this subject. In the beginning of this section I said the executive summary should be written by the proposal manager in collaboration with the marketing specialist. This, of course, is assuming that one or both of them knows how to write. The trouble with most executive summaries is they are written by amateurs. That is, one of the top management people or the president thinks, "Gee, I haven't really contributed anything to this proposal yet, so I gotta do something so that I can say *I* wrote the proposal just in case we should win it." I have yet to see a company executive who didn't think he could write. Without exception, their concept of writing is to cut and paste a conglomeration of mostly unrelated paragraphs that someone else wrote, then do a little wordsmithing with it, and presto, up comes a work of art!

If there is anywhere that the proposal calls for skillful writing, it is the executive summary, because this is where you get a chance to make a sales pitch, and good sales pitches call for skillful writing. People whose writing experience has been limited to dictating memos to the troops about unauthorized use of the duplicating machines just can't expect to be able to write skillful sales pitches. But the old management syndrome ("I can do anything better than anybody else except the president, because *I* am the vice president") takes over. So they grab the scissors and a roll of cellophane tape, and soon you have what any other amateur would come up with: a long sleep-inducing history of the XYZ Corporation and everything it has achieved since it was founded in Podunk, in 1873. I keep telling these people over and over and over again that nobody (that counts) is going to read beyond that first paragraph that tells where, when, and how the XYZ Corporation was founded. They don't give a damn about all that. They are looking for something they can relate to on *this* specific job you are proposing to do.

Summary: Executive Summary

1. It is, in essence, a sales pitch.
2. It contains no technical detail.
3. It is a brief summary of highlights—salient features.
4. General format:
   a. Critical areas for performance of contract
   b. Salient points of technical proposal
   c. Brief résumé of your company's experience and expertise in this work and why your company is best qualified
   d. Brief description of what is in the proposal
5. If not the most important, certainly the most visible part of the proposal.

In conclusion, your executive summary should answer the following questions, preferably in this order:

1. What are the critical areas for overall performance of *this* contract?
2. How do you propose to overcome or respond to these critical areas?
3. What are the unique, innovative, significant features of your technical or management approach?
4. Why are you best qualified to perform this contract?
5. How is your proposal organized?

TECHNICAL PROPOSAL:
THE THREE ESSENTIAL ENTITIES

If your specialty has been writing executive summaries and you now embark upon a technical proposal, you will have to shift gears. The technical portion is just the opposite of the Executive Summary in that there is:

No sales pitch

No generalities

No overriding need for brevity

Also, there is very little boiler plate to be found in the technical proposal. As a corollary of this statement, there is very little opportunity to use any cut and paste here. What the customer wants to know is *specifically* what are you going to do to solve his unique problem. Paragraphs lifted from some other proposal or from a tutorial paper cannot possibly be specific solutions to unique problems of a particular contract. If the statements so lifted can apply to two or more situations, they must, by definition, be generalities, and the evaluator will spot this artifice every time and will resent it because it insults his intelligence. Remember, the writing level here is keyed to technical staff people, most likely engineers, and administrators, thoroughly familiar with the contract requirements.

The technical proposal may address a myriad of technical subjects, depending on the nature of the contract. The RFP (instructions and statement of work) will indicate the subjects to be addressed and their sequence. *You* must decide whether additional items need to be discussed in order for your response to be complete. The subjects of the technical approach may include anything from systems concept to training; task organization and staffing to work flow, cost controls, and hardware/software interfaces. But the total extent of the technical proposal must cover essentially three major entities: nature of the problem, understanding the problem, and solution of the problem. The success of your treatment of any major section will be a measure of the extent that you address these three fundamental factors, the three essential entities:

Nature of the Problem

The nature of the problem may be addressed by restating the mission to be accomplished and then describing the features or special considerations that make this job different from others. (No two jobs are exactly alike.) *Do not* play back the RFP. Sometimes, for very complex missions, it may be necessary to paraphrase or even borrow a phrase from the RFP for precision of language, but then go on from there with a more detailed description of the mission. Be sure you have stated the mission precisely, because what you say here is especially sensitive when it comes to contractual liabilities under the ensuing contract.

Understanding the Problem
(Understanding the Requirement)

The best way to show your understanding of the problem is to point out the critical areas involved in the problem. For example, if the contract involves maintenance of delicate equipment exposed to a tropical, coastal atmosphere, this is a critical area and will require special attention. If the contract involves the design and procurement of a state-of-the-art equipment to be installed and integrated in remote overseas locations, this involves critical areas, and you must show the customer that you are fully aware of their magnitude.

Many RFPs actually require you to list what you regard to be the critical areas for successful performance. Whether it is actually spelled out or not, it is an important element in evaluating any technical proposal, guaranteed to be heavily weighted. After all, how can you solve a problem if you don't understand what the problem is?

It behooves you to use some serious in-depth thinking and imagination in identifying critical areas. Then, when you think you have analyzed it correctly, you must have some brainstorming sessions with other proposal team members, or your friends, to test the validity of your analysis and to ascertain whether you have touched all the bases.

Solution of the Problem

This is often addressed in the RFP as "technical approach." In any case, this is where you tell the customer what you are going to do to meet his requirements, how you are going to solve his problems—in specific terms, what you are going to deliver (labor, expertise, materials, equipment, and so forth) to meet the requirements of the RFP, and when, where, how, why, and by whom. You also point out the specific advantages of your approach over possible alternatives and also in terms of what it will do for the customer. Your technical expertise is the only limitation here.

A word of caution on responding to an RFP: The explicit terms set out in the RFP are what you must respond to and nothing else. You must interpret the language in the RFP literally. Again, *do not* get the idea you know better than the customer what he wants. What *you* think he *should* want is irrelevant. *Do not ever* propose something other than what he has specified (except separately in an alternate proposal). If you do, you will be adjudged nonresponsive (besides irritating the customer intensely). Finally, do not propose something extra that would be "nice to have" if it is going to cost anything. You won't get any extra points for it, and you will just be pricing yourself out of the competition. Again, respond to the RFP and nothing else, analyze and interpret the RFP literally, literally, literally. . . .

Normally the technical proposal (technical approach) is the most important segment of the entire effort. After all, this is the place where you, the prospective contractor *prove* that you know how to do the job. And that is really the whole *raison d'être* for the procurement process. Most executives I know think cost is the only factor. Nonsense! If cost were the only factor, the customer would simply issue an IFB (invitation for bid). Cost is always a factor, of course, but rarely ever the most important factor.

Only amateurs (and there are a few around) would award a contract for technical services or technical production based on cost alone, or even use cost as the main criterion. Why? Because, as in anything else, you get what you pay for, and I know a few cases here and there where the customer got himself into scandalous situations by awarding to the low bidder when any fool should have known the contractor could not do the job at his proposed cost. Both the customer and the contractor suffered in the process while the whole business community was laughing at their well-deserved predicament. Sooner or later it ceases to be a laughing matter, because either the customer or the contractor has to resort to desperate measures to save the situation (in the customer's case, it is known as saving face). This often leads to cheating and cover-up, subterfuge and chicanery, and eventually litigation (and in the case of government contracts—Congressional investigations and/or criminal indictments). One has only to read the daily newspapers to see instances of what I'm talking about. It happens all the time. The bottom line is: Don't think you can "buy in" to a contract. Both the contractor and the customer usually get exactly what they deserve—lots of trouble.

Now that I have put cost in its proper perspective, let's get back to the technical approach segment of the proposal. It is almost always the most important part of the proposal.

I once managed the *technical* proposal for a large contract where the technical portion was the *only* part that was even evaluated by the customer. This was most fortunate for us. Because of poor preparation, the *management* proposal was thrown together in a state of chaos and was, as to be expected, a mish-mash of uncorrelated cut-and-paste jobs; the phase-in plan was an unspeakable disaster; and even the executive summary was a rather lugubrious collection of redundant themes and little else (written, of course, by one of the chief executive officers—one of those with an attitude of "I know more about anything than anybody else, and I can do anything better, because I *am the boss*"). Even the cover letter was mostly good for laughs. It included an error in spelling that indicated gross ignorance of common English usage. In this case, however, the customer was so dissatisfied with the incumbent's technical performance that he was interested only in the technical capability of the bidders. Consequently, we won on the *technical* proposal alone.

How does one manage to write a better technical proposal than the contractor who has been there doing the job for 10 to 15 years? Whenever I suggest this is possible (at preproposal conferences, as I did in the above real-life case), I almost get laughed out of the room. Most people just do not understand that you, the outsider, have certain distinct advantages over the incumbent contractor. These advantages will be described in detail in the next section, Management Proposal.

However, suffice it to say here that you as an outsider can propose radical new ideas with much more freedom than the incumbent. Big remunerative contracts are seldom won by any company unless it proposes something new that enables the customer to get the work done more efficiently, at less cost, and with, it is to be hoped, less anxiety to the customer.

Let's take an example of an innovative idea to illustrate my point. Suppose you are bidding on a contract to provide transmission of precision radar tracking video data from various sites for display in a central control tower. Everybody else will probably bid a microwave system for the data transmission. Your brilliant, innovative engineers assure you that you can bid a video data compression technique that has never been done before, but it enables you to transmit the data over existing land lines at enormously reduced cost, as well as a great savings in time required for completion of the contract. Great idea. But you must prove to the customer's satisfaction in your proposal that this is feasible, and you know how to do it—and then prove it again with an oral presentation at the Q&A session.

Summary: Technical Proposal

1. Writing level here is keyed to technical staff members.
   a. *No* sales pitch
   b. *No* generalities
   c. *No* boiler plate

2. Usually will contain at least three major subsections:
   a. Nature of the problem
   b. Understanding the requirements
   c. Technical approach—solution of the problem
      This section should contain:

System concepts	Work flow
Interfaces	Cost controls
Quality assurance provisions	Trade-offs

Control techniques	Schedules
Task descriptions	Task organization and staffing
Training	Work breakdown structure

The extent will almost always be dictated by the RFP.

MANAGEMENT PROPOSAL: ITS OBJECTIVE AND CONTENT

The objective of the management proposal is to convince the customer that your company's management techniques, the quality of your personnel, the logic of your organization, and your related experience can best provide the planning, implementation, quality control, supervision, and support required to effectively perform the contract within schedule and budget constraints.

Whether it be hardware contracts (where hardware is delivered as the end product) or services type (including software) contracts, the principles of proposal writing are the same. However, in Government contracting there are basically two different types of technical service contracts: the mission type and the task-order type. The successful management proposal writer will carefully distinguish between these two basic doctrines of management and all the consequent subtle ramifications involved in management technique.

The essential difference between operating under a task-order contract and a mission-type contract is the extent of Government (or customer) surveillance. Under a task-order-type contract, the customer's monitors will exercise day-to-day surveillance of contractor progress on each documented task comprising the contract. Often, because of this close surveillance and lack of management flexibility by the contractor, such contracts take on the aspects of a "body shop" operation— an illegal personal services contract. The task order approach has been a favorite technique in NASA and has frequently gotten them in trouble with contractors and labor unions because of this aspect.

The trend now seems to be away from the task-order- and toward the mission-type contract in NASA (as elsewhere), due in part to the aforementioned troubles, but more so because of the tightening of funding for NASA programs. The same trend will undoubtedly spread to other Government agencies (including Defense) because of the inexorable pressure to decrease Government spending, along with a concomitant intensification of scrutiny into Government waste and Government procurement and management procedures. Obviously, the switch to mission-type contracts will enable the Government to reduce personnel ceilings and shift the management burden to the contractor at greatly reduced cost. The reason for this is that the Government can thus eliminate that skeleton force of monitors who simply comprise a dual management and engineering structure in any Government contracted project. Unquestionably, a greater and more demanding management burden is placed upon the contractor, but any well-managed company should be capable of absorbing the increased responsibility at minimum cost.

What is the nature of this increased burden on the contractor? Under a mission contract, Government direction is held to an absolute minimum. The contractor is given a general statement of objectives and goals at the inception of the contract; in other words, his "mission." During the course of the contract, the contractor may be given additional missions beyond the purview of the original mission; those will constitute modifications to the original contract, increasing the number of personnel and funding for the contract.

The contractor's manager must anticipate tasks required to fulfill the mission and schedule and prioritize his resources for accomplishing the mission in a timely manner. The contractor does not wait for any directives, schedules, task orders, verbal orders, budgetary allocations, staffing ceilings, or any other type of direction from the customer. His only concern is that the contract objectives be achieved, that they be achieved on time, and that the total expenditure at the end of the fiscal year does not exceed the contract limitations. Of course, the contractor is obligated to keep the customer informed of progress, problems, and plans. This is accomplished through formal periodic reporting systems, contract data requirement list (CDRL), and also informally through the customer's department heads.

It can be readily seen that this approach imposes a much greater responsibility on contractor management, calling for much greater initiative, imagination, planning, and manpower and budgetary control than in the task-order system. The only way the customer is going to get in the act is when it becomes necessary to forestall a crisis situation. There is no one looking over the shoulders of the working troops telling them when they are about to do something wrong or questioning their wisdom in taking such and such an approach. If you make a mistake, it will show up on your performance evaluation. In short, no one is going to tell you *how* to do anything, but you will hear about it if it turns out wrong.

Of course, the customer does maintain surveillance, but at upper-management level. The striking difference is that you, the contractor, tell him what you are going to do to meet contract requirements, not the other way around; then the customer approves or disapproves. The means of keeping the customer informed of status will vary from place to place. One way is to submit a five-year plan for approval and update it every year. Then you prepare a briefing, project by project, for carrying out the major goals set out in the five-year plan.

The CDRL should cover all necessary formal reporting such as a monthly report of contract costs, broken down into manhours; overtime hours; salaries by labor grade; materials expended; equipment procured; and so forth.

Management personnel must keep their counterparts informed, but again, the emphasis is on *telling them* what you are proposing to do, not asking for guidance. If they disagree with the outline of your approach, it is up to them to advise, but the ultimate responsibility for success or failure is on the contractor.

The customer can show his displeasure with a contractor by (1) informal verbal advice, (2) performance ratings, (3) contract action, that is, monetary penalties, termination, and so forth. Or, the customer can show approval by the opposite: praise, contract renewal, award fees, and so forth.

In other words, you must show that you know how to do all the things the customer either does for you, or forces on you, in a task-order contract.

In addressing a mission-type contract, you must show the Government that you know:

- How to budget time, money, manpower, and other resources
- How to generate internal controls and records
- How to issue specific directives, set goals, allocate resources
- How to establish surveillance and feedback systems
- How to anticipate and respond to emergencies or unforeseen circumstances

On the other hand, the task-order contract in its most austere form (some would say, most pernicious form) is analogous to the blanket-ordering agreement familiar to the commercial world. The customer provides a detailed specification of exactly what he wants accomplished to fulfill each task; the contractor determines the cost of time and materials, describing the skill mix required, and any support requirements to be provided by the customer. Then, a firm fixed price is negotiated between the contractor and the customer. Sometimes the contractor even has to provide résumés of those who will do the job and accept penalties for failure to deliver in the agreed time frame. Obviously, the contractor assumes all the risk in this situation and the customer none. In the other situation (mission contract) almost the reverse is true. The prudent contractor therefore adjusts his bid accordingly. If the contractor must assume all the risk, the customer must pay a price for it.

I have described these two concepts in some detail, because the philosophy of management and therefore the management approach either case are almost the antithesis of each other. This has a significant impact on the selection of the program manager and the techniques applied in preparing the management section.

As with the technical proposal, the management proposal should follow the general format: problem, understanding of the problem, solution of the problem. Start out by stating the management mission—the essence of successful management of this contract. Then discuss the critical areas in managing *this* contract. If, for instance, the contract involves management of a number of small installations dispersed over a large geographic area, you must recognize and acknowledge that this is a critical area. If it involves the servicing of a large number of diverse, scattered users, or if the nature of the operation is subject to notoriously high attrition rates in personnel, these are critical management areas and you need to let the customer know you recognize them.

Finally, you must provide realistic, practical, and specific means of solving these problems. For example, let us say it involves servicing a large number of users in the management of an ADP (Automatic Data Processing) system.

You might propose:

- Periodic distribution of a newsletter to all users incorporating acceptable ideas for improvements, new instructions, and so forth.
- News flashes on the terminals;
- Operations analysts providing training and consultation for users;
- Daily systems stabilization meetings to investigate causes of any malfunctions and issue action items for correction and prevention;
- Rigorous configuration control procedures.

Any self-respecting management proposal should, at a minimum, address the following issues:

1. Description of the type of management the company intends to provide for this contract; for example, centralized or decentralized control; autonomous, semiauto-

nomous, or corporate control (like who and under what kind of limitation makes binding decisions affecting the contract).

2. Position of the program manager in the corporate organization. (Does he have access to top corporate management?)

3. Limits of authority and responsibility of subordinate managers and supervisors.

4. Proposed organizational structure showing lines of authority. (Obviously, organization charts are required here.)

5. Interfaces by all levels of management with customer, user, associated agencies.

6. Management control techniques. How do various levels of management monitor work progress, quality of performance, adherence to established policies, procedures, techniques, configuration management, performance assurance, reporting procedures, quality control, and the like? In other words, "feedback systems."

7. Ways and means of staffing the contract, recruiting methods, retention of incumbents, promotion and other employment incentives, in-house training, leadership, and so forth.

8. Caliber and experience of management personnel. Brief one-paragraph résumés of key management personnel in text.

9. Management techniques for training, motivating, and guiding personnel (management training courses, technical in-house training, cross-training, promotion from within, awards, bonuses, promotions, morale and *esprit de corps*).

10. Charts. The management plan must have many charts. Use them to break up the monotony of pages of type by interspersing them in the text (not as separate pages). The following are considered imperative:

    Organization charts
    Function charts
    Interface charts

Work flow charts
Schedules
Staffing
Phase-in milestone chart
In some cases, even PERT charts

The management plan is commonly a catch-all for all supporting activities of a program, such as safety, security, personnel and administrative policies, procurement policies, and résumés, company-related experience, corporate finances, and so forth.

There are many areas in the management proposal where you can cut and paste your way through, for example, safety and security, but *do not* try it on the basics: the problem, understanding, and solution, because you can't respond to specific requirements with generalities.

You, as an outsider, have many advantages over the incumbent, as I have already pointed out. This is generally even more true in the management proposal than in the technical proposal.

I know of several contracts, each of which was valued in the hundreds of millions of dollars, that changed hands in just the past few years. In every case, the incumbent had become firmly entrenched over a period of 15, 20, or even 30 years. How can such things happen? How can an outsider possibly come in and write a better proposal than the people who have lived with this thing for all those years? The answer is that it is one of the greatest paradoxes of the real world that the outsider has some very distinct and often overwhelming advantages over the long-term incumbent. That is what I am going to tell you about in the remainder of this chapter.

HOW YOU CAN BENEFIT BY NOT BEING THE INCUMBENT CONTRACTOR

I have seen many people, charged with responsibility for preparing a management proposal, get discouraged before they start. They go around wringing their hands and wailing in anguish,

"How can we possibly write a better management proposal than the incumbent who has been there managing this operation for ten years?"

Well, let's take a good look at the situation.

1. Perspective

The longer a contractor has been on a particular contract, the more he loses his perspective. The incumbent gets to where he can't see the forest for the trees. He gets careless, complacent, over-confident, and sloppy. And he becomes more and more oblivious to the necessity for making the big cost sacrifices that a hungry competitor will make. There is a tendency to develop tunnel vision and thus miss the handwriting on the wall that forecasts the need for drastic changes. There is a tendency to write a proposal that tells the customer what one is doing rather than what one should be doing to meet the challenges of future years. There is a tendency to reject innovative ideas and radical cost-saving devices, because "We have always done it this way" or "What we are doing has worked, why change?"

If you have been managing a contract for 10 to 15 years and getting reasonably good performance ratings, you are naturally reluctant to make drastic changes. For one thing, you apply the old adage of engineers, "If it ain't broke, don't fix it." Trouble is, your 95 percent rating might get compared to what is, in effect, a 115 percent rating if some imaginative bidder with special expertise comes along and proves he can do the job more expeditiously, with less cost, and with greater future growth potential—growth in efficiency, growth in higher applied technology, and accelerated cost savings.

There is a tendency on the part of every incumbent to defend the status quo, and worst of all, to think he knows more about what the customer should have than the customer does.

There is a tendency even to become arrogant to the point of thinking you are running things, and the customer is, in effect, working for you. I know of a case in point. All employees were allowed to pass through the guard gate (operated by the incumbent contractor) on a military base for a one-hour lunch period. The guards kept track of the time and reported tardiness to the culprit's supervisor. They noted that the young military officers frequently used up more than an hour, because they did their jogging to keep in physical condition as required by the military and then had their lunch. So, the contractor indignantly reported these infractions to the base commander, demanding that officers be held to the same rules as employees. That was the beginning of the end for that contractor.

You may question the veracity of the above tenets, but I assure you I have seen those things happen over and over again in more than 20 years of observation.

I helped manage a contract proposal once where the incumbent program manager was the proposal manager, which is as it should be. The biggest trouble was, the customer (NASA) wanted a transition from a work-order-type contract to a mission-type contract. Nine times out of ten, the incumbent simply cannot make such a radical change in philosophy. He just cannot help writing down just the way he has been doing these things for the past five to ten years. I would review and reject a proposal contribution, because it was still describing a work-order mode of operations. Then I had to take on both the contributor and the proposal manager. The argument I usually got, both from the program/proposal manager and the contributor was: "I just don't believe they really mean to go completely mission oriented on this contract. Why, I heard Bill Shmo, the PCO, say just the other day . . ." "But I'm telling you, dammit, you have to go by what the written word here in the RFP says, not by what Bill Shmo says." "But, he's the purchasing contracting officer. . . ." "I don't care if he's the Pope; the written word is what you have to go by. Did you ever hear of a verbal contract for $65 million? And besides the final award decision is made by higher authority than the PCO." But it was a losing battle. It was like pulling teeth to get them to remove even a fraction of that labyrinthian structure of reporting to counterparts at every level down to technician—a wasteful and unnecessary exercise, an example of bureaucracy gone beserk, but so beloved to the old school of NASA-trained bureaucrats. Needless to say we lost the contract. It was a shame too. It was one of the best *work-order-type* proposals I ever saw.

The tendencies are the consequence of natural human failings potentially existing in all of us, whether we be managers, executives, supervisors, or whatever. That is why the best managed corporations rotate the management of their contracts every three to four years to prevent this "fat cat" attitude from developing. Here is an axiom to paste in your hat. If you have a "fat cat" manager, you have a "fat cat" organization. Bear this in mind: You, the hungry competitor, have a significant advantage in perspective over the incumbent.

2. Freedom to Change

The incumbent contractor is often placed in a terrible dilemma—a catch-22 situation when recompetition time arrives. If he or she doesn't make radical changes, he or she loses the contract. If the incumbent does make radical changes, the customer may say, "Why did you idiots wait until now to propose all these great ideas? You have been ripping us off for years, you crooks!" So you, the hungry competitor, have the freedom to come up with refreshing, cost-saving, clever, innovative ideas—the incumbent does not.

What the incumbent has is an operation that evolved through experience, through necessity, through trial and error, and any new fundamental concepts have been dismissed as too radical and disruptive, also risky. Everyone is coasting along, fat, dumb, and happy: "Don't make waves, sonny; you haven't been here long enough to understand."

The only limitation is, of course, that your clever ideas have to make good sense. You have to make a case for them in your proposal *and* be prepared to defend them in the orals—the Q&A session that precedes every major contract award.

Let's consider a management-oriented bold innovation. You know that an auto manufacturing company has been procuring his supply of widgets from a New England-based firm for 30 years. You know that this firm has an outmoded plant, that it is practically run by union bosses with a featherbedding mentality. You know that the laws of that state, in effect, subsidize strikes by paying unemployment benefits to strikers, that taxes are unbearable, and that his prices are exorbitant because of these problems. You can buy a large facility at low cost in a right-to-work state in the South. The auto manufacturer can put up the money for equipping the plant with state-of-the-art equipment using new sophisticated manufacturing techniques. You prove that he can amortize these initial costs over five years and then continue to buy your widgets for half his current cost. Actually, that is not so innovative after all, because it has been going on for years now.

You could propose something that your customer might find refreshingly new. It's called leadership. I once saw a very wise old lady being interviewed on television. She had pioneered many of the techniques involved in advancing the computer industry to its present state of the art. Near the conclusion of the interview, she was asked, "How did we go wrong? How could this great country—once unchallenged by the industrial world—become threatened and outwitted by people in a little group of islands in the Far East, a country that we had annihilated in 1945?"

Her reply was something for us all to ponder and reflect upon, "We have had too many *managers* and not enough *leaders.* You can manage commodities, material, supplies, facilities, and business, but you don't manage people. You must *lead* people. This requires leadership." What a magnificent idea! Why didn't the "managers" of our industrial giants and their counterparts in government think of that as they collaborated in turning over management of business to the parochial and short-sighted interests of political pressure groups and powerful labor unions? Because they were "managing," not leading, because they were maximizing profits instead of providing courageous leadership; because they were looking toward the next quarterly financial report instead of the financial health of the company ten years hence; because they were planning how to make a bundle and retire, and to hell with everything else! The welfare of the company or the industry or the country ten years from now would be a problem for their successors. These were "managers," the product of Academe—of courses in economic theory, charts, graphs, and abstractions; the obedient dupes of liberal doctrinaire professors who never had to meet a payroll or joust with an obstreperous union negotiator.

There appears to be a new breed of leaders emerging from the brink of our near economic disaster. Recently I read of one who had built a $90-million business in a few years starting from a stake of $600. He hires a new crop of college graduates every year to become the future managers, directors, and vice presidents of his still rapidly growing business. What he tells them at an annual meeting is this: "I suppose many of you are already carving out a secure niche for yourselves in this company, jockeying for position to move up the corporate ladder, and applying the theories of management you learned in school. And as soon as I find out who you are, I am going to fire every last one of you." I guess what I'm trying to say is: Our colleges have taught our people too much management theory and not any leadership at all.

The lack of leadership is accountable for the poor judgment so frequently used by managers in personnel matters. Alas! The government must share a good deal of the blame for personnel mismanagement since they have taken it upon themselves more and more in recent years to intervene in day-to-day personnel management decisions. And with the usual disastrous results encountered when the Government intervenes in business management. (Of course, the Government would never have had the opportunity to get in the act if managers had been doing their job fairly and intelligently.)

Back to incumbent weaknesses: Nine times out of ten, the incumbent does not see the overall situation in a realistic perspective, does not have the freedom to make drastic changes, does not use objectivity in staffing levels, and cannot be objective in assignment of key personnel.

Another thing that handicaps incumbents is the old "promotion to one's level of incompetence" syndrome (usually referred to proudly as "promotion from within"). Good ol'

Joe Zilch is a good super-tech, with 20 years' experience (that is, one year of experience times 20), who knows XYQ-24 radars from A to Z. So he gets promoted to managing a whole group of sensor systems: TM, photo-optical instrumentation, acoustical sensing arrays. Too bad he never had some academic training for understanding these multidiscipline systems and the principles of physics that make them work. He has been promoted to his level of incompetence and the incumbent is stuck with him and the customer knows it.

The incumbent's freedom to replace Joe Zilch and objectivity in dealing with the problem is hampered by the knowledge that the customer is going to ask how come you put him in there in the first place if he was not competent to handle it: "What kind of stupid management have you got?"

The reason for this contractor's difficulty in the first place is the time-worn policy of "promoting from within" that, with the help of labor union mentalities, has achieved the same level of unassailability as motherhood. Of course, this is good policy provided one has management material from within to promote. But, like any other policy, this one must be implemented with good, common sense, and keep sentiment out of it. It is always a good policy if implemented with sound management decisions, but who ever said that managers always make sound management decisions?

The same thing applies to staffing. An incumbent has been milking the contract for years by overstaffing. After all, it is a cost-plus-fee contract. So, how can he suddenly decide he can do the same job with 20 percent fewer people now that the contract is being competed?

He has several good ol' boys in key slots. The head of procurement has been surreptitiously scrounging unauthorized use of the customer's property for the program manager for years. His reward has been promotion to a labor grade far in excess of that authorized by the contract or justified by his responsibilities. Is the incumbent going to brazen it out by proposing him in that labor grade, or reduce him to the proper grade (which might take some explanations besides the chance he might vindictively blow the whistle on those nefarious activities)?

Furthermore, there are some skeletons in the closet that the incumbent has been successful in concealing up to now. How about unauthorized use of the customer's computer, reporting of overtime not used, charging his contract with work done on another (losing) contract; unauthorized attribution to acts of God to justify late performance on the contract. I could go on and on. I've seen them all. All you have to do is read the papers. Now you, the proposal manager of the hungry competitor, must subtly weave these needles into your proposal, *not* by direct reference, but by *proving* how your organization has built-in safeguards and enlightened management to prevent these things from happening.

Where do you get this information on the incumbent? From your marketeers. And if they are too shy and lacking in initiative to ferret out this information, *fire them all* and start over. Otherwise you are working in the blind. I knew one consultant marketing specialist who hadn't even heard of the competitor who eventually won the contract. If you want to talk to him, I think he is driving a cab in the Bronx now.

Another thing that *good* marketeers should strive to obtain is the Independent Government Cost Estimate (IGCE) for the program. This, of course, is not released to the public. Don't ask me how to get it. If I knew I'd probably be a marketeer—an affluent one. That's why you select marketeers with initiative, imagination, brass, unscrupulousness, and charm—expecially charm. It helps immensely if they belong to a country club, drink in the right bars, are always seen with the best people (VIPs), have good ears, good memories, and can hold their liquor.

I once worked on a contract that, if we had known the IGCE, we would have won hands down. We had the technical expertise, had a presence in the area, and had a good technical proposal, but the CEO insisted we low-ball it constantly. We came in with practically a skeleton organization at about half the IGCE. The winner's proposal wasn't all that good, but he knew the IGCE and accordingly bid just under it. They concluded accordingly that he understood the contract better than we.

Summary: Management Proposal

1. Objective: To convince the customer that your company's expertise, organization, personnel, and experience can best provide the planning, leadership, quality assurance, and supervision required to perform within schedule and budget constraints.

2. Should contain:
   a. Description of the type management your company intends to provide
   b. Position of program manager in corporate organization
   c. Limits of authority and responsibility of managers/supervisors
   d. Proposed organizational structure showing lines of authority
   e. Management control techniques-feedback systems
   f. Ways and means of staffing the contract
   g. Program control to assure timely performance of contract objectives
   h. Caliber and experience of management personnel
   i. Management techniques for training, motivating, and guiding personnel.

3. The incumbent has many handicaps to overcome:
   a. Perspective; hard to see the big picture; innovations
   b. Restricted in the ability to make drastic changes
   c. Difficult to be completely objective in staffing and retention of key personnel

PHASE-IN PLAN: THE DOWNFALL OF THE UNWARY

Although the phase-in plan is often included in the management proposal, it is addressed separately here because of its singular importance. Some imprudent proposal managers treat it almost as an afterthought, assigning it to some dud on the proposal team just to get him or her out of the way. Such a manager must be totally lacking in empathy. Put yourself in the customer's shoes and imagine yourself one of the customer's department chiefs: (of course, you as department chief have been assigned to the Evaluation Board to review the proposals). You have been working with the incumbent contractor for five years and although you know he is not perfect (you even have to chew him out once in a while), he is still a *known* quantity, and he at least has familiarity with the job. Furthermore, you have one or two critical ongoing projects and if either of them gets botched up, heads will roll, probably yours. You've read contractor X's technical proposal and he seems to know how to do the job all right, so now you turn to the phase-in plan.

Well, he jumps right in and says he is going to have the program manager and technical director report on Day 1. But wait, what are they going to do? They don't say they are going to do anything except meet the contracting officer. Are they going to recruit and process the 65 incumbent technicians themselves, or are they also going to bring in some personnel administrators to do this? He doesn't say. He says he is going to bring in key personnel early on. But he doesn't say what these key personnel are going to do. You have to conclude that he doesn't know what they are going to do. So having brought in the key personnel he starts talking about his company's history in retaining incumbent personnel. You don't give a damn at this point what he has done in the past. You want to know what he is going to do here and now, and *how* he is going to do it. Then he goes on to tell you how important the phase-in is, and blah, blah, blah, *ad nauseam*. At this point, there is *no way* contractor X is going to get this contract—only over your dead body! Believe it or not, I have seen proposals like this and even worse.

Of course, what the customer department chief wants to know are specifics, not these empty platitudes. He wants to get a nice, warm feeling that you are capable of planning this phase-in period so flawlessly that there will not be so much as a hiccup when you take over.

What the customer wants to know is: What is your plan for phasing over responsibility for various operational areas and/or shifts; for assuming responsibility for customer owned property, equipment, and facilities; for initiating coordination and orientation between key customer personnel with their counterparts in the contractor organization; for training and orientation of personnel, and so forth. You, the proposal writer, simply must project yourself into the new environment and live there until you finish writing this section.

Here are other examples of items the customer may want to hear about:

Recruitment and transfer of incumbent personnel

Sources of incoming personnel (company transfers, new hires)

Inventory of property

Coordination with labor unions

Arranging security clearances

Payroll and accounting procedures

Leases, bank credit, logistics

And so on

As with any time-phased activity addressed in a proposal, this discussion *must* be illustrated on a milestone chart showing the sequence, duration, and schedule of all major events involved in the phase-in cycle. See Figure 3-1 for an example of a milestone chart for a 60-day phase-in.

On this example we have a contract award notice just two days before the beginning of phase-in (not unusual). The program manager does not run off half-cocked and unprepared to meet the customer's management and contracting people immediately. He telephones the customer and arranges a meeting four days later. This gives him an opportunity to convene his phase-in team and review the proposal phase-in plan. (It may have been turned in months ago.) Specific dates and schedules will be set up for accomplishing specific goals during this meeting; travel reservations will be made; appointments will be set up with counterparts of the members of the phase-in team; there will be review of the proposal, dissemination of information to key personnel who will be assigned to the contract, and on and on with a myriad of details that must be planned and coordinated before reporting to the customer's facility.

FIGURE 3-1 Milestone Chart: ABC Contract

	D-Day	D+10	D+20	D+30	D+40	D+50	D+60
Contract Award Notice	▽						
Initial Meetings:							
Phase-In Team	▽						
Program Manager-Customer	▽—▽						
Arrival Phase-in Team	▽						
Personnel Director-Customer	▽	▽					
Finance Director-Customer	▽	▽					
Operations Director-Customer	▽	▽					
Engineering Manager-Customer		▽—▽					
Logistics Manager-Customer		▽—▽					
Negotiate Labor Contracts		▽———▽					
Prepare MIS Systems		▽————▽					
Review and Revise Maintenance Plan		▽————————▽					
Review and Revise Safety Plan		▽————————▽					
Review and Revise Security Plan		▽————————▽					
Finalize Employee Benefits Plan			▽				
Organize Configuration Control Board			▽				
Finalize Quality Assurance Procedures				▽			
Submit All Plans for Approval					▽		
Assignment of Key Personnel (Overlap)					▽———▽	▽	
Process Incumbent Personnel					▽——————▽		
Hire and Process New Personnel						▽———▽	
Transfer and Hire Date All Personnel							▽
Assume Total Responsibility For Contract							▽

While the details of these activities are being finalized, the program manager travels to the customer facility and meets with his counterparts in the customer organization. He goes over the phase-in schedule he submitted in the proposal plan and gets approval or makes necessary changes recommended by the customer. He gathers all the useful information he can during the next two days: ground rules, security requirements, availability of transportation, arrangements to orient incumbent personnel on what to expect, peculiarities of the local community, labor relations, housing situation, attitude of the local populace, and so forth.

On arrival of his phase-in team (five days after contract award date, seven days after award notice referring to the milestone chart), the program manager is fully prepared to brief the team on the current situation, give them a firm schedule for achieving their goals, and introduce them to their counterparts in the customer organization.

While his staff is coordinating with their counterparts on the detailed plans for phase-in of the various elements of the contract, the program manager has plenty of other things to keep him busy. He should take his operations director with him and make a complete tour of the sites or facilities involved in the contract, establish working relations with supporting sites or facilities and with parallel organizations whose support and cooperation he needs for fulfilling his responsibilities on the contract.

He must be aware of the fact that the people who will be supporting him are a part of the community. Sometimes the local community is hostile to newcomers; sometimes it is fearful that all its people are going to lose their jobs to these "outsiders." The program manager must allay these apprehensions and through personal contact with local leaders convince them that there will be no disruption to their way of life or their economy. It is mandatory that he make this reassurance, because the support of the local people is vital to his success in smoothly phasing in the contract.

He continues to meet with the customer's key personnel at all levels and establishes a high visibility, always with the purpose of presenting the best possible image of his company and presenting a cordial, businesslike, competent, no-nonsense, goal oriented impression of the new management team.

Concurrently, he schedules meetings between members of his phase-in team and the incumbent key personnel for the purpose of establishing a general *modus operandi* for the phase-in period. In particular he must establish a tentative agreement for phase-over of property accountability, transfer of fringe benefits, overlap of key personnel, transfer of software, O&M procedures, plant in-place drawings, coordination of replacement of MIS (Management Information Systems), and negotiation of any labor union contracts.

As you can see from the milestone chart, there are many activities going on, both in sequence and concurrently. To integrate all these interrelated functions into a coherent effort, resulting in the creation of a cohesive labor force takes *organization* and *planning* on the part of the contractor. And that is what the customer is looking for in the way you present your phase-in plan, *organization and planning*.

Generally speaking, you should take over all of the operational responsibility on a given day, not in phases or segments. The reason: It complicates matters for the customer by dividing responsibility if you don't. He wants either one contractor or the other in complete charge at all times.

To keep costs down, you should assign only the minimum number of your key personnel to overlap with the incumbent's key personnel, and for the minimum time necessary. More than two weeks is usually counterproductive. It is not uncommon for the bidder to propose absorbing phase-in costs in order to be more competitive with the incumbent (who has no phase-in costs). This really gives the program manager an incentive to hold down phase-in costs. As with the proposal effort itself, the phase-in effort can realize the greatest economy by timely and thorough planning and coordination. If you have prepared a good phase-in plan for your proposal, it not only impresses the customer, it saves your company dollars when the time comes to implement the plan.

Neither the milestone chart presented in this section, nor the outline for phase-in plans presented on the following pages is intended to be a panacea for everyone. They are simply presented as a pattern to get you started, a base line from which to proceed in preparing your proposal. Of course, you must adapt your outline to fit your own needs and the dictates of your special situation.

As indicated in the title, the phase-in plan is often the downfall of the unwary. *Do not* underestimate the importance of the phase-in section. Many proposals have failed solely because this section did not give the customer confidence that the bidder could accomplish the transition without creating an unholy mess that would disrupt ongoing operations and be an ignominious embarrassment to everyone involved.

SAMPLE PHASE-IN PLAN OUTLINE

INTRODUCTION: General statement of your company's policy and experience on phase-in, addressing incumbent retention on a selective basis, replacement of incumbent key personnel in policy-making positions, and program manager's role in creating a dynamic new work force in your company's image.

1.0 Perception of Critical Areas
    1.1 Management
        Replacement of key personnel
        Replacement of MIS systems (for example)
        Property accountability transfer
        Orientation of transferred incumbents
        Role of program manager in integrating this particular contract effort
    1.2 Technical
        Familiarization of your key personnel with customer's environment
        Transfer of technical control of operations
        Preparations for assumption of control without disruption of operations
        Alternative actions considered in effecting phase-in
        (*Note*: The above paragraph must be limited to a discussion of your *perception* of these problems, *not* the solution to the problems. The idea is to first persuade the evaluator we understand the major problems. Detailed solutions will come later.)

2.0 Staffing Plan

    2.1 Sources of Contract Staffing

        Show estimated numbers of:

            Incumbent personnel retained

            Your company's assigned personnel

            New hires

        By the following categories:

            Management

            Technical areas

            Support areas

    2.2 Time Phasing

        Depict the above numbers versus time on a matrix, broken down by job classification.

    2.3 Critical Skills

        Identify and describe how you will provide a contingency plan for responding to the possibility of certain portions of the work force not available by contract start date.

3.0 Phase-in Preparation

    3.1 Organization of Phase-in Team (or Cadre)

        Provide organization chart

        Provide milestone chart of phase-in team assignment

    3.2 Duties and Responsibilities of Phase-in Team

        3.2.1 Program manager

            List individuals he will contact and *specifically* what matters he will discuss with each. For example, the program manager visits the customer's contract manager and discusses his policy in regard to interfacing with contractors, any potential problems or uncertainties regarding logistic (or other) support, critical time periods anticipated by the customer in fulfillment of his mission, and so forth.

        3.2.2 Operation's manager

            Same approach

        3.2.3 Other managers

        (*Note*: The list of items to be discussed should collectively address every detail that needs to be resolved and every item of information that is needed in order to effect a smooth and orderly takeover of contract responsibility by a thoroughly oriented, fully informed, qualified phase-in team. For example, the program support manager with (among others) the incumbent logistics manager discusses specific plans for inventory of GFP (Government Furnished Property) and the mechanics of assumption of responsibility for property accountability. This kind of detailed list of people to contact, and *specifically* what items are discussed, must be accomplished for each member of the phase-in team.)

4.0 Policy on Interfacing with Incumbent
Include an interface chart here to show who interfaces with whom. Outline procedures for resolving disputes, with resolution by the customer being only the last resort. Describe how phase-in team will work out interface details during initial conferences with incumbent (preceding paragraph). Program manager discusses incumbent's phase-out obligations *in his initial meeting* with the Contracting Office.

5.0 Assumption of Transferable Personnel Liabilities
(Continuation of fringe benefits for retained incumbents. This is where you describe procedures for processing transfer of incumbent hires.) (Your Personnel and Finance Department provide this.)

6.0 Security Clearances and Assumption of Accountability for Classified Documents and Material
Briefly describe the procedures and objectives you propose to employ for this purpose.

7.0 Phase-in Procedure
7.1 Detailed Method for Transfer of Responsibilities
7.1.1 Work performance
Task acceptance and accomplishment
7.1.2 Work documentation
Task control
(Method of processing and assigning tasks during phase-in.)
7.1.3 Status and cost reporting

7.1.4 Engineering; fabrication and installation, drafting and technical literature

7.1.5 Configuration control
Documentation

7.1.6 Operations control

7.1.7 Project management

7.1.8 Project administration
Include—
Property inventory and signover
Payroll and accounting procedures
Personnel records including clearances
Transfer of responsibility for existing subcontracts, if any
Initiation of commercial relationships with suppliers (of spare parts, materials, transportation, POL [Petroleum, Oil, Lubricants], and the like)

8.0 Conclusion

A short sales pitch to reassure the evaluator you understand the magnitude and complexity of the problem of phasing into *this specific contract* (skip the generalities, the platitudes, and the clichés cut and pasted from old proposals) and that you are well prepared and experienced enough to handle it *without any interruptions* of ongoing activities.

CHAPTER FOUR

GOOD WRITING TECHNIQUES— AND HOW TO GET STARTED

THE GOLDEN RULE OF PROPOSAL WRITING

The golden rule of proposal writing is: "Make it as easy as possible for the reader to understand and evaluate your proposal." Most of the cardinal rules that follow are but corollaries of this golden rule. If your proposal is well organized so that the evaluator can find what he is looking for, and if you write authoritatively in a crisp, straightforward manner with narrative that flows in a logical sequence, you have gone a long way toward writing a winning proposal. You might say that the only remaining ingredient is to come up with the right solutions to the customer's problems.

THE CARDINAL RULES OF PROPOSAL WRITING

1. First and perhaps most important among the cardinal rules is: "Respond fully and precisely to *all* RFP requirements." It is truly amazing how many people simply ignore significant portions of the RFP. I don't think it is just plain sloppy preparation that causes this, because I've noticed some otherwise thorough people do it. The more likely explanation might be that many people *think* they know better than the customer what he wants. Therefore, they simply omit certain responses because *they decide for themselves* that certain instructions or SOW (Scope of Work or Statement of Work) items are just too trivial to bother with. I assure you that every such omission will constitute a deficiency—too many deficiencies and your proposal ends up in the wastebasket. So follow all RFP instructions

99

to the letter and that means *all* instructions. Sometimes the instructions will not all be in one place. They may pop up from time to time throughout the RFP. That is why it is imperative that the proposal manager go through the entire RFP and mark these additional instructions.

2. "Follow the format of the RFP as much as possible." Your proposal format should be a sort of mirror image of the RFP. Wherever the RFP instructions are not explicit in format, use the format that most closely matches that of the RFP (including sequence). Of course, if the RFP or statement of work rambles, repeats itself, is disorganized, confusing, and virtually incomprehensible (as many of them are), then you have a problem. You have to present your work in a logical and orderly manner. The best way to do this is make a list of all the RFP (or SOW) requirements and then sort these requirements into logical groups like maintenance, planning, quality control, and so on, and then organize your presentation in the most logical manner that addresses all these requirements. You may have to use some plain common sense here.

3. "Be positive, specific, to the point, and straightforward." Try to sound honest—none of this used car salesman stuff. Don't try to evade or "paper over" an issue. They will see through you every time.

Say you are bidding on a hardware contract to provide a communications system. You know that it would appear desirable to use a fibre optics link as part of the system. But you don't know if the customer would want to go the fibre optics route. (That's because you haven't done your homework.) So, you straddle the issue. You propose a microwave system. "Of course, if a fibre optics link is desired here we could provide . . ." Then you go on to tell them how smart you are at fibre optics. Result: Another contractor analyzes the alternatives and makes a convincing case for a fibre optics link and makes a positive decision himself and proposes accordingly. He knows what is best. You don't. He is in. You are out in the cold.

4. "Tell them what they want to know and no more." If someone asks you "What time is it?" they don't want you to

tell them how to build a clock. Also, they don't want a free textbook course in anything. So lay off the tutorial papers. If they want that, they will go to the library, not put out an RFP.

5. "Make the format pleasing to the eye." Nobody wants to look at page after page of monotonous, soporific type. Break up the monotony with judicious spacing or use key points (bullets) to emphasize important points. Whenever possible, intersperse the text with charts, tables, diagrams, or pictures. *Do not* put these graphics on separate pages if at all possible.

6. "Make your narrative flow"—from the general to the specific, from the known to the unknown, from the past to the present to the future, from the simple to the complex. Don't get cute or try to titillate them with obscurities, or test their intelligence with esoterica. Remember, you are not trying to make the world forget James Joyce. You are simply trying to communicate some fairly simple concepts to some fairly simple—make that intelligent—proposal evaluators.

7. Address the Three Essential Entities described in Chapter 3 to wit: (1) state the problem (statement of the mission, (2) state your understanding of the problem (identify the critical areas for successful performance), and (3) set forth your solution to the problem (*how* you are going to accomplish the job). The articulation of these three entities is indispensable to every major section of the technical proposal and management proposal.

8. "Put yourself in the evaluator's shoes." "Empathize"— this is really just a corollary of the Golden Rule. You cannot write effectively unless you can project yourself into the evaluator's shoes and tell him all about what you are going to do.

THE MORTAL SINS OF PROPOSAL WRITING

1. "Regurgitating the RFP." Almost everyone tries this the first time he works on a proposal, thinking perhaps no one will notice. How naive! This transparent substitute for intelligent

response can only incur the wrath of the evaluator and do irreparable harm to the whole proposal. In-house review of the proposal must concentrate on eliminating every vestige of this practice. There is only one occasion when you might be justified in borrowing just a phrase from the RFP and that is when you have a very complex mission to state and you need to be extremely precise in order not to give the impression you are attempting to modify the mission in any way.

2. "Obfuscating the problem." This is a favorite of tricky lawyers who have a weak case. The idea is to get the jury so confused that they declare the defendant not guilty so they can all go home. The evaluators will see through this smoke screen every time, however, and they will simply give you the minimum grade and move on to the next proposal so they can go home.

3. "The cutting-and-pasting syndrome." This is the most insidious evil of all, because it sometimes passes for intelligent thought, that is, to everyone but the evaluator. The really important parts of your proposal cannot be cut and pasted. Why? Because those important parts demand specific answers to specific requirements involving unique situations. Obviously something that was written about another situation cannot be applied effectively and convincingly to your situation. Just changing the name of the players won't hack it. They will see through this cheap little ploy every time. This is not to discourage the use of old proposals as reference materials. In fact, you haven't done your homework if you haven't researched old proposals on similar contracts before you start to write. But, they are to be considered background to provide ideas; they might touch on some points you overlooked. Now use this information, but apply your own thinking to the specific set of circumstances at hand. There are proper places to cut and paste. No sense in re-inventing wheels. The "company-related experience" section, the résumés, the "company administrative policies and procedures," safety plans, security procedures—just make sure you revise them to conform to RFP instructions.

4. "Vague, empty generalities." This is a close relative of the cut-and-paste syndrome. I once had some idiot complain to me

that one of my proposals "wasn't any good." (It was good enough to win the contract.) Why wasn't it any good? Because he couldn't find anything in it he could cut and paste into any proposal he was working on. I considered this a real, if inadvertent, compliment. That means that I was writing specific answers to unique problems, not a conglomeration of vague, empty generalities such as he was accustomed to using. Come on, folks! You are talking to real people, about a real, specific, finite, living, breathing problem. Quantity is not the criterion; relevance is.

5. "Passive voice and negative statements." The passive voice is the favorite mode of writing for bureaucrats. This is because the passive voice always leaves one in the dark as to *exactly who* does something. That way you can't pin the responsibility on anyone when something goes wrong. In the passive voice the doer of the act is vague and usually purposely so. Examples:

Richard Nixon: "The tapes were somehow erased." (Wrong)

"I erased the ** [expletive deleted] tapes myself. After all they were *my* tapes." (Correct)

Jimmy Carter: "It was decided to use only six helicopters for the rescue mission in Iran." (Wrong)

"The stupid ** (expletive deleted) Joint Chiefs of Staff assured me that six helicopters would suffice." (Correct)

Take the positive approach always. Example:

"Absenteeism will not be tolerated." (Wrong)

"The ABC Company director will take the following steps to eliminate absenteeism. . . ." (Correct)

6. "Rambling—going around in circles." This is the product of a disorganized mind. If you have a rambler on your proposal team, identify him early and banish him from the premises. Such people are dangerous, and their problem is usually incurable. They are dangerous because the evaluator can go crazy

finding the particular section he is supposed to be evaluating. This is because he will read something and say to himself, "This must be where they are addressing hardware interfaces." Then he happens to see something a few pages further on and he says, "No, I guess *this* is where they are addressing hardware interfaces," and so on and on. This is likely to make him very unhappy and frustrated. Need I say more? Rule: *Do not go on to a new subject until you have thoroughly and completely covered the old.*

7. "Bare, unsupported statements." This is a common failing among proposal writers, even experienced ones, and especially where they are the incumbent. It is natural to think, subconsciously, "The customer already knows this, so I don't need to elaborate on it." Trouble is that oftentimes the evaluation board may be made up partially of outsiders who don't know you from Adam—and they are not intimately familiar with the details of your operation. But even if they are familiar with the operation, the use of bare unsupported statements is never enough. Remember, you are responding to an RFP, to the printed word—not to certain individuals. Example: "Our past nine years' experience on the XYZ contract has given us a deep insight into the needs of the ABC Agency. . . ." And so much for understanding the problem. "Well, hey, I'm the customer and you may not have gained any insight at all in nine years. Prove it to me." Just saying so doesn't prove anything. So remember this: A bare unsupported statement can only be considered an opening gambit. You've got to take it from there and prove your premise is correct.

8. "Long-winded tutorial lectures." This is a first cousin of the obfuscation problem. Sometimes a proposal writer will not know just how to address the customer's problems or even to identify his problems. So he thinks, "I'll overwhelm them with my erudition." He copies half a textbook on the subject, or maybe throws in the company SOP plus a couple of magazine articles he saw recently. *Don't do it.* They will spot you for the charlatan you are. Besides, no one wants to read all that irrelevant hogwash.

9. "Incessant, redundant bragging." The braggart is another character you must spot as early as possible because some of them are incorrigible, and drastic action is imperative. The best way of handling them is to channel their energy into socially acceptable pursuits such as company related experience, résumés, or gathering material for the executive summary. If you have ascertained that they are incorrigible, keep them away from the management proposal and the technical proposal at all costs. As emphasized before, the only place you are allowed a little bragging is in the executive summary, and then you must do it with restraint, discretion, and good taste. If you overdo it, it becomes counterproductive. So, perhaps you can use the braggart, but keep an eye on him.

10. "Superfluous, verbose, prolix, tautological redundancy." Not to mention repetitiousness. This is a first cousin to the rambler. These people (known as motor mouths) sometimes labor under the delusion that the evaluator uses a bathroom scale to evaluate your proposal. Some others among them don't need an excuse. It just comes naturally. One reassuring thought about the rambler, the tutor, and the motor mouth: They can usually be spotted during the outline phase of the proposal (because these kinds of people do not have the self-discipline and organizational ability to prepare an acceptable outline). This gives you time to take defensive action and prepare to neutralize them or eliminate them.

A List of Don'ts

- Don't start sentences with a prepositional or adverbial phrase. Example: "With the renewed interest in environmental protection programs and plans to upgrade the sensor systems for measurement of atmospheric state, and considering the sophisticated advances in the state of art for gathering of such quantified data, the XYZ Corporation proposes blah, blah, blah." (Wrong) People like me with normal size brains just can't carry all that long-winded baggage in memory until the writer gets to the point. "The XYZ Corporation proposes . . . We feel that the renewed interest in environmental . . ." (Correct)

- Don't harp on past experience. Too much turns people off.
- Don't repeat yourself or return to a subject you've already covered.
- Don't use excessive verbiage. Lengthiness is not a criterion of quality.
- Don't make unsupported, dogmatic statements.
- Don't submit busy-busy charts, diagrams, and the like.
- Don't preach or teach. Spare them the tutoring.
- Don't use vague generalities. You are talking to real people about a real-life, specific, problem.
- Don't string 15 adjectives together; for example, "The data obtained from the servo-driven, low-light-level, recording automatic, digital, optical, laser-ranging and tracking device. . . ."
- Don't try to snow the reader. He might know more than you think.
- Don't use the imperative "shall." You are writing a proposal, not a specification.
- Don't present important decisions, solutions, conclusions, or declarations without rationale to support them.
- Don't use expressions like and/or, his/her, their/our, office/work area. Better to use more words than to sprinkle your writing with these train-of-thought stoppers.
- Don't just tell what you are going to do. They also want to know how, why, and by whom.
- Don't start writing until you have made an outline of your approach.
- Don't start writing until you have reviewed the pertinent instructions from both the RFP and the proposal manager.
- Don't pick up your pencil until you have thought the problem through.
- Don't try to think the problem through until you have researched the problem.

- Don't try to research the problem until you understand what the problem is.

- Don't think you understand the problem until you have discussed it with other competent people.

And now for a few pet peeves: There are some words and expressions that just make me grind my teeth, and I can go through almost any proposal and pick them out like ants in a picnic lunch or pebbles in my shoes. Here are a few (but not necessarily all) of the expressions I despise. I just used one of them—the expression in parentheses in the previous sentence. Of all the abrasive expressions in the English language, this has to be near the top. Nobody wants to see weasel words in a proposal, not even the bureaucrats who undoubtedly invented that expression. They want to see forthright, upbeat statements, not equivocations.

Here are some of the other (ugh) expressions:

1. "Cognizant" as in, "The cognizant engineer will ensure that all safety precautions are taken. . . ." Well, who in blazes is the *cognizant* engineer? That is just the point. The writer didn't know. Maybe it's the radar supervisor, maybe it's the QC engineer, maybe it's the operations manager, or ??? So he takes the easy way out and says, "the cognizant engineer," which tells the reader he didn't know.

2. "Philosophy" as in "Our management philosophy is blah, blah, blah, zzzzz." Leave philosophy to the philosophers. You are dealing with realities, facts, decisions, numbers—not philosophy.

3. "Adherence" as in "Supervisors will ensure adherence to established procedures." Fine, that's what they are expected to do. Would anyone propose otherwise? Any fool can "adhere" to established procedures. And how do you "adhere"? What are the "established procedures"?

4. "Highly qualified" as in, "Our highly qualified engineers will prepare design specifications. . . ." Really! Is anyone else

going to propose just run-of-the-mill engineers who are *not* highly qualified? There are proper areas of the proposal where you can, and *must* show how you are going to provide "highly qualified" engineers, and so on, *and* what these qualifications will be. But when you go on saying, "Our highly qualified, specially trained, greatly motivated, superlatively reliable, unbelievably versatile, extraordinarily educated (usually sober) engineers and technicians will perform blah, blah, blah," you are just making yourself sound ridiculous.

5. "Facets" as in "Our experience covers all facets of the contract. . . ." I must confess I am not sure precisely why this word bugs me. But it is probably because it is overused, really doesn't say anything, and it seems the writer usually thinks, having used a cliché like this, that he has said it all. Why did he not use the word "aspect," or "phase"? Because those words have the connotation that you are going to delve into the details thereof. But "facet," according to the dictionary, means "polished, plane surfaces of a cut gem," and that's about all you ever get when you see that word. You never really get the gem.

6. "Existing procedures" as in, "Maintenance will be accomplished [ugh, passive voice too] applying existing procedures ... blah, blah, blah." This is another example of those glittering generalities such as you usually find with "facet, philosophy, and cognizant." The writer obviously doesn't even know what the existing procedures are.

I could write out a few more examples but I'm beginning to feel ill. This should be enough to give you the idea, anyhow.

HOW TO GET STARTED WRITING

OK. You have done all your preliminary study, analyzed the RFP, constructed your detailed outline, and know all about cardinal rules and mortal sins of proposal writing. *How do you begin writing?* First of all, there are no panaceas; it is hard to set out any hard and fast rules that apply to every proposal. Every RFP is unique. Regardless of any suggestions I put forth here, remember that they are superseded by any RFP instructions that may be in conflict.

There are, however, a few basic, simple, and of necessity, very general rules for getting started.

Rule 1: Set the stage.

Have you ever noticed when you go to a movie, it usually opens with an exterior scene (often while the titles are still superimposed on the screen)? Let us say you are the director and the dialogue begins with an attorney and his client discussing a case. Now, if the director started his first scene there, you would be asking yourself, "*Who* are these clowns? *Where* are we? *What* are they talking about? *Why* is this one guy talking about a murder?" And if you didn't get some answers, pretty soon you probably ask yourself, "*What* am I doing here?" and get up and go home.

In order to avoid all this confusion, the director shoots what they call establishing shots. The first scene may show a street with heavy traffic, sidewalks crowded with well-dressed pedestrians walking briskly past and in and out of adjoining buildings. The next scene shows a man emerging from the other pedestrians, carrying a briefcase, entering one of the high-rise buildings and getting into the elevator. Next scene shows him emerging from the elevator, entering an office with a name on the door and "Attorney-at-Law" underneath. He waves casually to the secretary as he enters an inner office. Without a word being spoken, you know his name and that he is an attorney whose office is downtown in a big city.

This is what you have to do in "setting the stage." You do not want your evaluator saying to himself, "who ...? where ...? what ...? or he (the evaluator) may end up saying to himself, "What am I doing here?" while tossing your proposal in the wastebasket so he can go home.

Don't jump into the middle of a subject thinking you will come back and pick up the beginning later. Remember the cardinal rule for narrative that flows: State the mission, the objective of the contract, as you understand it. You must get these who, where, what, why, when questions answered early; then you can get into those fascinating details or those clever technical innovations you have in mind.

Rule 2: Show that you understand the problem.

Nearly every important subject that you have to address in the management or technical proposal will have a few critical areas that are central to an understanding of this particular section of the contract. You must identify these critical areas, because, most important, it shows you understand the problem, but also because it gives your narrative a focus, something the evaluator can relate to, something that will pique his interest and hold it— something that will make him say, "Ah, here at last is a contractor who understands our problems; now let's see if he knows what to do about them." Now, does anyone think he can accomplish this with vaguely relevant abstractions that were cut and pasted from some other proposal? Or some gratuitous platitudes about what a great company you are and how experienced you are at solving such problems? No way. And yet, show me almost any old proposal (especially losers) and I guarantee that I can find some examples of this sort of thing written by people who thought they could pass it off as intelligent thought.

Rule 3: Show that you know what to do about solving the problem.

You have stated the mission. You have identified the critical areas and shown that you understand the problem. Now show that you know what to do about it; what you are going to do to fulfill the mission, perform the work, minimize the risks, react to the inevitable mishap, respond to the potential obstruction. Now here is where some proposal writers get all fouled up. The object here is to show *how* you are going to solve the problems involved in the contract, indicate the technical and management approaches intended, what will have to be done, and of course how, when, where, and why. But you do *not* have to actually solve all contractual problems in your proposal. Sometimes proposal writers will get sidetracked onto this endless trail and find themselves inevitably bogged down in a bottomless morass of detail from which they cannot extricate themselves, because neither time nor space permits. Remember: the ultimate solution of the problem is what you are *contracting* to do, not what you are expected to provide in the proposal.

What is the *level of detail* to be described in response to the statement of work? That has been the most difficult concept for most technical people I know. The tendency with those most intimately acquainted with the subject is to get into things like spray painting of electrical wiring to inhibit corrosion, instead of talking about how you develop policies and procedures for overall preventive maintenance of the project. Of course, I can't give you any hard and fast rules to cover all situations, but here is a rule of thumb that covers most:

> You must address the statement of work to a *level of detail one magnitude greater* than what is set out therein.

For example:

> SOW: Conduct pre-mission checks and calibrations.

You must describe what checks and calibrations will be performed, by whom, and when. How do you validate system configuration? How do you verify readiness to support the mission? Pretend you are the unit supervisor here and you are briefing some VIPs at your site. You don't know how much they know about your system, but you have to assume they know something (maybe more than you do) or they wouldn't be there. You tell them how this system works and how *you* make it work. You don't tell them how you spray the electrical wiring and all that. They haven't time to listen to that.

All of this book, up to this point, has been concerned with the mechanics of preparing for and writing the proposal or segments thereof. It has been suggested to me, both in the courses I have presented and in writing this book, that I include some good and bad examples of actual proposals. While this would indeed be instructive, it is quite impractical. Even the smallest of proposals would run some 50 pages and the reader would be bored to death, especially reading the bad ones. Since this is the last chapter in the *mechanics* of good proposal preparation, however, I think it would be appropriate to include some horrible examples of bad proposal writing—just excerpts that will illustrate some of the points I've been preaching. This will be followed by a critique I once wrote of a *bad* proposal where the poor old XYZ Corporation didn't do anything right. I believe these real-world examples will serve to demonstrate the kinds of mistakes and sloppy writing that should be avoided as well as serve as a review of many of the points I have been making up to now. I have underlined the most horrible phrases for your convenience.

Incidentally, you may have noticed throughout this book that the ABC Corporation does everything right, and the XYZ Corporation is a loser; they do just about everything wrong.

Example 1 (Of Bad Proposal Writing)

Having received early committments *(sic)* from a   (1) (4)
sufficient quantity of incumbent employees, our   (3)
phase-in team will turn their *(sic)* attention to: (2)

- Becoming familiar with the detailed workload   (3)
  requirements as they exist at that point in   (3)
  time;
- Establishing themselves as co-managers of,   (5)
  and co-contributors to, the existing task
  requirements.

Comment: Here we have in one short paragraph an incredible example of sloppy work: (Numbers in right margin correspond to numbered paragraphs following.)

(1) An obviously misspelled word—not just a typo.

(2) A mistake in grammar that any fifth grade student should spot.

(3) Vague, indefinable generalities that mean absolutely nothing. (I underlined these phrases.) *What* is a sufficient quantity? *Who* in our phase-in team? What detailed work load requirements? And how on earth is the reader supposed to determine what these detailed work load requirements are or what are "the existing task requirements"? And also, when is "at that point in time"?

(4) Beginning a sentence with an adverbial phrase. Of course this is not a hard-and-fast rule for all cases, but for proposals one should, for the most part, stick to simple declaratory sentences. This helps give the crisp, businesslike tone that helps create a favorable image of your company.

(5) And finally, any fool should know there is no such thing as co-management.

Example 2

The interface with ABC Corporation is expected to be somewhat sensitive; therefore, XYZ will conduct their interface with the objective of receiving maximum possible support from ABC. XYZ cannot interfere with responsibilities of ABC for their (sic) residual(?) contract responsibilities; neither can XYZ tolerate lack of cooperation by ABC. . . .

Comment: Don't you just know this is going to make the customer feel warm and happy and good all over, confident that you are going to handle the interface problems with tact, diplomacy, and smooth efficiency?

Example 3

XYZ is attuned to the issues and has an exceptional history of successful performance in contracts requiring disciplines similar to the present procurement. (Underlined to point out sloppy, imprecise language.)

We have developed and refined a detailed methodology that will ensure a smooth performance of contractual responsibilities. We rely principally upon the assignment of key personnel, our ability to foster a strong coordinative effort between both client and user personnel. The approach and specific methodology contained herein is based upon experience we have gained. . . .

Comment: Zzz . . . z. What did he say?? Answer: absolutely nothing. Had enough? Well, just one more short one.

Example 4

The subject of change to the procedures and mode of operation may represent a critical issue if it is not treated appropriately. XYZ approaches this issue very simply. We do not change procedures or modes of operation. . . .

Comment: Obviously these guys really did a lot of homework on this one to come up with such a profound solution.

And now for a critique of another real-world example of sloppy proposal writing. You will note there are many errors brought out here that I have discussed in this chapter. (*Note*: The following section is an actual evaluation of a proposal; all references pertain to material within the evaluation.)

EVALUATION OF XYZ DATA CENTER PROPOSAL

1.0 Introduction

This evaluation was conducted for the sole purpose of identifying weaknesses in this proposal as a means of pointing the way to improvements on our next effort. I have gone through and analyzed the entire proposal, section by section, and have tried to make constructive suggestions for improving our next effort. I am sorry to say that it is *not* difficult to understand why this proposal did not favorably impress the customer.

2.0 Executive Summary

This entire section looks like a cut-and-paste job. Almost every sentence could have been written for almost any facilities management proposal. There are no interesting ideas and almost no specific solutions to problems. In fact, there is very little recognition of what exactly are the critical areas involved in this contract. The entire section is so laden with vague generalities, stilted clichés, and stereotyped sentences I doubt any evaluator would read it to the end.

The executive summary should be a good sales pitch. As such, it should, above all, get the reader's attention and hold it. This is accomplished by giving him confidence that, "Here at last is a contractor who understands our problems."

The executive summary should start out with a brief non-technical summary of the customer's problems or requirement with special emphasis on what are regarded as the critical areas of successful performance. This is to show that you understand the problem.

This should be followed immediately by *specific* measures XYZ will employ to take care of the customer's needs. Give the customer some meat to chew on here. Don't save the goodies for elsewhere in the proposal, because this is probably the only part of the proposal that the Evaluation Board chairman or the top management people will ever read. Make every word count. Don't waste space with generalities that can only turn the reader off. For example, under "Hardware Maintenance," in telling how XYZ is going to handle the maintenance subcontractor, it says, "XYZ's approach is to exercise vigilance in monitoring and controlling the maintenance subcontractor. Anything less than a rigorous watch would lead to maintainability and reliability deterioration." And so much for hardware maintenance and subcontractor surveillance. How profound!

What the customer wants to know is, *Who* is going to do what to whom, and *how* is he going to do it, and *how* is management going to see that it gets done.

A few more observations on the executive summary:

Impact. In order to hold the reader's attention it should deal only with the *high points*, the basic decisions involved, the innovative ideas, the adroit, clever, original plans we can come up with. Certain sections could be introduced with a hard-hitting sales pitch sentence like (for the section on user interface) "XYZ currently responds to the needs of no less than

3500 clients from a cross-section of government, commercial, and industrial organizations throughout the world." I think it has much more impact to sprinkle these statements into appropriate places throughout the summary than to put them all in one paragraph.

The summary should be prepared *early* in the proposal development *to serve as a theme guide to all proposal contributors*. This obviously was not done here, because there is no detectable theme in this summary. Furthermore, if it had been done early, it would not have had so many typos. As stated before, this is the one section the customer's top management is likely to read, and numerous typos indicate sloppy performance. That is one impression you do not want to give. On two pages, I counted no fewer than ten typos.

Charts. Since only the highlights should be included in the summary, I would eliminate Figures 1-3 and 1-4, the latter being too busy and both contributing little that a higher level reviewer would be interested in. Both figures belong in the detailed management plan only.

The section on cost proposal (paragraph 1.7) like most of the rest of the summary says nothing of any substance.

In conclusion, the executive summary must be completely reorganized and rewritten from scratch. And this must be done early in the next proposal effort, so that this document can serve as the theme guide to all proposal contributors.

EVALUATION OF XYZ PROPOSAL (continued)

Section 2: Identification of Functions

The proposal instructions mandate seven sections plus a cost proposal as follows:

1. Executive Summary
2. Identification of Functions
3. Technical Approach to Functions

4. Management Approach
5. Contract Staffing
6. Implementation Plan
7. Corporate Experience

The instructions for Section 2 state simply "the vendor should *restate these (18) functions* and identify any additional functions that may be proposed." The 18 functions are listed elsewhere in the RFP, grouped into four categories: Operations, User Support, Application, and Program Management/Cost Control *in that order.* Your proposal adds two more functions: Configuration Management and Quality Assurance, which seems like a good idea.

I would conclude from the above that the Proposal Section 2 should be confined to "restating the 18 functions," that is, a statement of the mission setting forth the scope and breadth of these functions *as you understand* them. In other words you are expected to define the parameters of the work to be done without any discussion of how you are going to do it or who is going to do it. That is reserved for Section 3 "Technical Approach to Functions." The instructions for this section state, "the vendor should identify the technical approach to be used for each support function and demonstrate clearly how they will be integrated within the overall management structure."

In view of the above, I cannot perceive what rationale was used for organization of Section 2 of your proposal. It starts out with "Understanding the Mission Assignment," which, in my opinion, should have been included in the Executive Summary. Then, the next paragraph is "Understanding the Functional Requirements," which is all right except the evaluator is going to be asking himself, "Is this where he addresses 'Identification of Functions'?" (which is what we are supposed to be talking about here). Then the ensuing order of paragraphs has no relationship whatsoever to the order in which they are presented in the RFP. The evaluators must have gone out of their minds trying to evaluate this proposal. I can readily see why they told you, *"It took longer for us to evaluate this proposal than it did for you to write it."*

The content of each paragraph is also in conflict with the RFP instructions. Instead of "restating the functions," it gives us a smattering of management details and procedures; for example, "all system failures must be documented and a file maintained by the section manager for subsequent analysis."

The cardinal rule for proposal writing, to wit, "make it as easy as possible for the customer to evaluate" has been repeatedly and flagrantly violated here.

So much for Section 2.

Section 3: Technical Approach

The first mistake in this section is the title. The proposal should, insofar as possible in every detail, match up to the RFP. This is to make the evaluator's job easier. The title in the RFP is "Technical Approach *to Functions.*"

There is some good raw material in this section and obviously a lot of verbiage. It just does not, however, conform to the RFP and is therefore nonresponsive. For example, wouldn't it be better to make the organization conform to the grouping of functions as set out in the RFP (page 4 of 5)? That way, you have three operating elements and a management function with each operating element—Data Center Operations, User Support, and Applications—having exactly the same functions as set out in the RFP. It would appear this is the way the customer wants it organized, so why confuse him, or why antagonize him?

Again! *User Support* should be a major heading with a grouping of functions as listed in the RFP. But in the proposal we have, para. 3.2 User Assistance; para. 3.2.7 User Services Support; para. 3.2.8 Training Documentation and *User* Communication; para. 3.2.8.5 *User* Communications. There is some good material here on user support, but it should all be organized under *one major paragraph heading*, so the evaluator doesn't go crazy trying to find it.

The entire section could use some surgery by someone technically oriented. Some of the material looks as if it came out of a textbook. Also, there are still more examples of vague generalities, for example, systems programming. There just isn't enough *who* or *how* in this section. "XYZ will incorporate all modifications to generate a new or updated system. . . . etc." It isn't enough to say XYZ will do . . . , they want to know *who* in XYZ, and *how* will it be initiated, reviewed, checked, implemented, and so forth. What interfaces are involved with the customer and how are they handled? Always bear in mind we are supposed to be writing to specific people about a specific, discrete facility, not about theoretical places or academic problems.

Some good points:

- There is much good material in this section that can be used *as a starting point* for the next proposal.
- All charts and tables are informative and look good except for the Weekly Keypunch Report, which is unintelligible. No one is going to have the patience to pick his way through that labyrinthian mess. I think the other charts would be more useful if they included the "who" responsible for various steps or phases.

Suggestion: We should print the applicable portion of the table of contents on each divider page, making it easier for the evaluator to find things.

Section 4: Management Approach

The instructions for this section state (inter alia), "The relationship between the contractor management and (customer) management *should be emphasized.* The vendor is encouraged to include flow diagrams that will depict the *decision-making process and lines of communication.*" One would conclude that he is especially interested in seeing these things that he has taken the trouble to specify. Yet I see no paragraphs in this section devoted to customer interface. I see no diagrams devoted to customer interface, nor to lines of communication, nor to decision-making process. True, there is a "Project Work Flow Integration" diagram, a "Work Cycle Overview," and a

"Work Flow Control" diagram, but *this is not what the customer requested*. These charts are fine as far as they go, but they do *not* respond to the RFP. The customer would like to know, for instance, how we interface with the Software Review Council, how we interface with the Production Control office, the failure analysis function leader, or at least that we know we are *supposed* to interface with them. Also, how do we interface with the 2500 users? It is not enough to say that so-and-so interfaces with them. There has to be a systematic interfacing process, a *modus operandi* for interfacing.

There should be a more painstaking delineation of corporate support. True, this is an autonomous contract, but the corporation provides guidance and assistance in many ways: recruiting, reassignment of key personnel to the contract, financial backing, legal counsel, industrial relations, and assistance in negotiating labor agreements, and so forth.

In the discussion of each major organizational element from the president of XYZ down to the key personnel of the project organization, there should be a thumbnail sketch of the background of these people—just two or three sentences. It makes the proposal more interesting to read and gives us a chance to brag a little without causing offense.

All of the organizational elements, users services, operations, and so on, write-ups need to be beefed up, with emphasis on *how they are managed*, not just what they are supposed to do. This includes interfaces, coordination, control, feedback, relationship with program management, relationships with users, and the like. Generally speaking, this section is somewhat more salvageable than the others.

Section 5: Implementation Plan

Like much of the rest of the proposal, I think this section has some good material in it, but it needs to be tightened up with some ruthless editing. It is too long on generalities and too short on specifics. For example, it describes the assignment of key personnel (page 6-3) to arrive during the prephase-in period, but it doesn't say what any of them are going to do. The evaluator would have to conclude that we do not know what they are going to do. The director and technical director are brought in during prephase-in, but who is going to recruit and process the incumbent hires (some 65 to 70 people)—the director?

Having phased in these *two* personnel, we start talking about incumbent retention history and recruiting. The section should be reorganized so that we cover all this peripheral stuff first and then get down to brass tacks and tell them exactly what is going to happen on a day-to-day basis until we are ready to assume the contract. Like, how do we recruit incumbent and local personnel, orient XYZ personnel, phase over various specific operational areas and/or shifts, arrange for banking/payroll service locally or where, setting up of off-site office space, assuming responsibility for GFE and facilities, coordination with customer management, with incumbent contractor, with local authorities, and so on, and so on, and so on.

The whole section jumps back and forth from one subject to another and never does get down to the practical down-to-earth realities of what needs to be done for a successful and smooth phase-in.

In summary, this proposal generally:

- Is poorly organized
- Does not effectively respond to the RFP
- Is too verbose—requires extensive technical editing (263 pages is much too long for this type effort)
- Is long on generalities, short on specifics
- Has some good material that could serve as raw material for the next proposal
- Is very difficult for a proposal evaluator (customer) to check against the RFP

End of Critique

Summary of Chapter 4

1. Every RFP is unique.
2. Guidelines for starting
   a. set the stage.

      b. provide the reader an establishing shot.

      c. Show you understand the problem.

      d. Show you know what to do about the problem.

3. Remember the cardinal rule for narrative that flows: Go from the general to the specific, past to present, and so forth.

4. *Project yourself* into the customer's environment.

Example

1. *Set the stage.*

      a. Statement of the mission.

      b. Unit or section performing the mission

      c. Position of unit in organization (use organization chart)

2. *Show you understand the problem.*

      a. What you will do to solve the problem

         (1) Minimize the risk.

         (2) React to the inevitable mishap.

         (3) Respond to the potential obstruction.

CHAPTER FIVE

HOW THE CUSTOMER EVALUATES YOUR PROPOSAL AND MAKES THE CONTRACT AWARD DECISION

First, we will present a general overview of proposal evaluation as seen from the customer's side. Terminology will differ somewhat among various customers or Government agencies, but the general principles are the same everywhere. This chapter is based primarily on standard government policies and procedures, but the material presented here can be as readily applied to any procurement agency of the government, or any company in the private sector that is procuring goods or services on behalf of the government, or any business making large scale procurements.

"Now why," you might ask, "do I need to know how the customer evaluates the proposal? I'm not in the procurement business." No, but you are going to go through a continuous process of evaluating your own proposal on the basis of what you think the customer wants to see or doesn't want to see. You must have a good insight into what thought processes the evaluator goes through and a thorough knowledge of what he is looking for, in order to respond adequately to the RFP. Remember: the sole purpose of writing a proposal is to convince a group of evaluators that your company is the most capable of performing this particular contract.

PROPOSAL EVALUATION BOARDS

The diagram (Figure 5-1) shows the functional structure of a Department of Defense (DOD) evaluation team. The Source Selection Authority (SSA) for large procurements is usually the senior military commander having an interest in the procurement if DOD; for other Government departments, a senior civil servant appointed by the department head. He makes the final decision for award. This person may overrule any recommendations of the Source Selection Advisory Committee (SSAC) or Source Selection Evaluation Board (SSEB), or may even refuse to accept recommendations. The SSA may request that their reports contain only the facts and findings. But remember, the decision must be supported by the documentation; that is, the SSA must be able to justify his decision in many cases to higher authority before it is final. He must always be prepared to justify the decision in case of a protest and possible GAO investigation. It should be noted here that, whereas protests are rarely successful (it takes a pretty flagrant case), the threat of a protest does help to keep the Government honest. That is why the proper implementation of their proposal evaluation procedures is so important. The system is designed to prevent fraud and favoritism by requiring documentation of all steps taken in arriving at the award decision.

The Source Selection Evaluation Board performs 95 percent of the work of evaluating the proposal and the negotiation of terms. The SSEB is usually (but not always) directed by the

FIGURE 5-1 Government Evaluation Structure

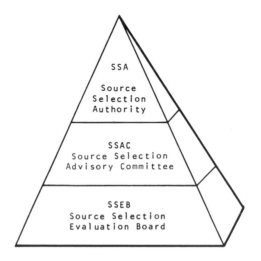

program manager for the program, that is, the user of the particular goods or services being procured. The people on the SSEB are the people who compare, in detail, your proposal versus the RFP. (Incidentally, no one really wants to be on the SSEB. As in assignments to write proposals, the SSEB members are usually technically experienced engineer-types, chosen from the field to perform this onerous task as an additional or temporary assignment.)

SOURCE SELECTION PROCESS

Usually the cost proposal is bound separately from the technical; if not, the proposal is submitted in loose-leaf binders and broken down into discrete sections before evaluation is commenced.

When SSEB evaluation is completed, those failing to score within competitive range limits (established by the SSEB before RFP release) are eliminated. (See Figure 5-2.) The Government is reluctant to eliminate competition, however, so final approval for elimination of a proposal must be secured from the SSA. If he feels that the deficiencies are such that they cannot be corrected, the proposal is eliminated. The Government is anxious to avoid elimination of a competitor, especially when there are a small number of bids submitted.

I know of one extreme case of this where the Government told the bidder he had failed to make the competitive range. The proposal manager, being an eager young fellow anxious to learn, asked the Government for a debriefing to find out how he had failed so miserably. So the contracting officer, being an affable and fatherly type of gentleman, patiently explained how he had misinterpreted the RFP here; failed to address a requirement there; failed to use the prescribed format for résumé, and so on. After he finished, he told the proposal manager there would be similar opportunities coming along later. As an afterthought, he mentioned that there were still three days left to resubmit his deficient proposal if he wanted to try. The proposal manager hastened back to his office, worked around the

FIGURE 5-2 Source Selection Process

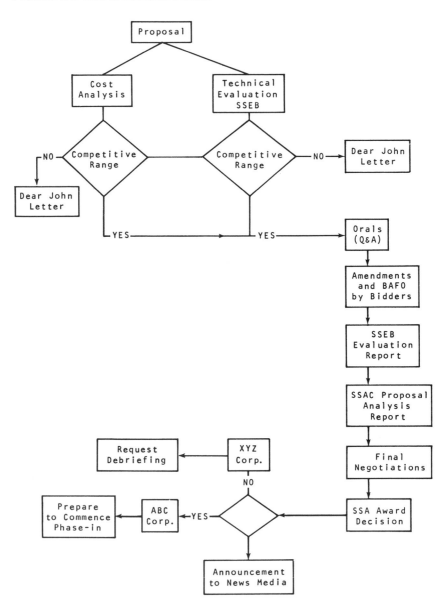

clock for three days, resubmitted, and won the contract! Of course, there were probably grounds for protest here, but no one protested.

Next is the "Q&A session" or "Orals." This is the only opportunity for the contractor to clear up any misunderstandings, correct any deficiencies, and fill in any inadvertent omissions. The Government's purpose is to give each qualifying contractor an equal opportunity to make such minor corrections that may have been the result of misinterpretation of the language in the RFP, simple inadvertence, or comparable matters that do not in essence bear upon the bidder's qualifications to perform the contract.

Most important is the fact that this is also an opportunity for the Government to question the key personnel bid on the contract, to see how they respond under fire, to test their knowledge of the contract in answering questions extemporaneously, and in general, size up the kind of people they might have working with them. This can be a very important phase of the procurement. In fact, it is often the make-or-break phase, where competition is close. It is therefore imperative that the proposal manager and his team do their homework before showing up for this meeting. For one thing, the key members of the proposal team and the key personnel proposed on the contract must become, and remain, thoroughly familiar with their respective portions of the proposal. This is the reason why it is generally recommended that the key personnel proposed on the contract actually be the key personnel on the proposal team. The best way to acquire the in-depth knowledge of the proposal, sufficient to survive the orals, is to have participated in the detailed aspects of the proposal preparation. Oftentimes the questions are submitted in advance, giving the contractor time to prepare with some deliberation. But the astute proposal manager prepares for the worst-case situation: questions directed orally and answers expected immediately.

Lastly, the people who show up at the orals must present a neat, businesslike appearance—suits and ties, no white socks, no beads on the men, no heavy makeup on the women. If you *must* bring someone along with a short fuse temper, give him some prior indoctrination on keeping cool.

The next (and final) step for the bidder is the BAFO (best and final offer), which gives the Government a final chance to squeeze the last drop of blood (if any) from the contractor before making the award. Oftentimes a bidder will hold out a little up to this point; then the Government thinks they have made a helluva deal when he shaves off a little more at this time (which he always intended to do). Also, it can work the other way; that is, the contractor can raise his price. I know of one case where the proposal manager (who also in that case did the costing—not a good idea) made a $300,000 error in the wrong direction. If the Government had not been so greedy as to ask for a BAFO and instead simply accepted his bid, they could have been into him for $300,000. But when they asked, he reviewed his figures, realized his mistake and raised the BAFO by $300,000. He didn't win the contractor, but then he wasn't out $300,000 either.

Now the SSEB renders its evaluation report to the Source Selection Advisory Committee (SSAC). On the basis of this report, the SSAC prepares the proposal analysis report (a somewhat condensed form of the SSEB report) and forwards it to the SSA for a decision. Usually, there will be some last-minute negotiations with the contractor selected for award. These final negotiations may involve a multitude of details that in the aggregate may amount to a significant amount of money. Such negotiations, however, cannot involve any issues that go to the heart of the contract—the essence thereof—otherwise a protest may ensue with much heartburn for everyone concerned. When these negotiations have been concluded, the contract is signed and the award is announced.

SSEB RESPONSIBILITIES

The SSEB is responsible for evaluating the proposal against pre-established criteria and standards. Where do these criteria come from? They *should* be prepared by the same people who are going to evaluate the proposal. This is to minimize any discrepancies in interpretation of criteria as between two different groups of people. These criteria (which appear in the RFP) are based mainly on the statement of work, the model contract found in most RFPs, and the proposal instructions. The "standards" do *not* appear in the RFP (by law). They comprise the detailed instructions by which the SSEB team evaluates your proposal. We will get into more on this later.

The SSEB establishes the competitive range, both on cost and technical score. The technical range, so established, is reviewed by the purchasing and contracting officer (PCO), the SSAC, and, frequently, also by the Source Selection Authority (SSA). This *should* be done before RFP release and certainly before proposal submission.

The competitive range is a more or less arbitrary minimum score under which your proposal will be eliminated from further consideration. This is especially important when a large number of proposals have been submitted. This enables the evaluation to focus on the three or four or five best proposals, and avoid wasting any more time on the "dogs." Government authorities, however, astutely avoid throwing a proposal out when it is close to the competitive range.

The weight of GAO decisions forces the goverment to keep the bidders in the competitive range rather than rigidly eliminating them. The controlling criterion is, "Can the proposal be corrected without major revision?"

The SSEB conducts meetings with each bidder individually to define contract terms, clarify obscure points, and correct inadvertent inconsistencies. Generally, only those bidders found to be in the competitive range are invited. Legally, a contractor may make changes right up to contract award, but the Government does not like it, and will probably find legal ways to frustrate such action. Finally, the SSEB prepares its evaluation report and briefs the SSAC.

EVALUATION PROCESS

Let us return to evaluation criteria. The establishment of evaluation criteria is an SSEB task, but must be approved by the

SSAC. These are the criteria you see in the RFP. There are three kinds of criteria: (1) those you see in the RFP; (2) those that only the Government evaluation team sees (called *standards*); and (3) those that no one sees, called biases, preferences, prejudices, emotional responses.

The first kind—those that you see in the RFP—are the ones that you must use as a guide in writing your proposal.

Following is an example of criteria as stipulated in an RFP for procurement of technical services to support a large missile range.

Example of Criteria for Technical Evaluation of Quote in Order of Importance (RFP)

1. Engineering and analysis

2. Management and support

3. Instrumentation operation and maintenance

4. Communications

   Within Each Functional Area

1. Approach
   Perception of problem areas and sound solutions
   Management and supervision
   Feasibility of technical approach to each task
   Extent of support resources (vehicles, housing, and so forth required)
   Degree of assurance that the bidder can actually perform
   Promotion of efficiency in accomplishment of tasks

2. Personnel
   Quantities and skills proposed; qualifications

3. Experience
   Recruitment for overseas operation
   Developmental engineering; procurement of complex equipment

This is just one of many ways the criteria may be listed. They could be described in narrative form, for instance, and as frequently happens, be given a comparative weighting to establish a ratio of relative importance. Remember, the "standards" (by which the Government evaluates) must be inherent in the broad criteria set out here. Nothing can be evaluated as a standard that is not intrinsic to the criteria set out in the RFP.

Before we move on, let's take a closer look at the customer's evaluation criteria as set out here. As I stated previously, the first cardinal rule for proposal writing is to "respond fully and precisely to all RFP requirements." Many times, the failure of proposal writers to respond fully and precisely to the RFP requirements is attributable to a failure to interpret the language of the RFP accurately. Generally speaking, RFPs are not written by talented or professional writers, and certainly not by lawyers. The language is consequently often obscure, imprecise, sloppy, poorly organized, and ineptly phrased. I have seen many RFPs that would take a Philadelphia lawyer to decipher. That is why I have emphasized that you must analyze, dissect, reread, and brainstorm the RFP (including SOW) before you even begin to think about responding to it. Sometimes it has crossed my mind that all this obscure language is used on purpose—to confuse the unwary and to separate the conscientious and analytical bidders from the sloppy and careless ones who don't deserve to win the award.

The example of criteria for technical evaluation is a case in point. It lists "criteria in order of importance." There follows a list of the major areas of the contract. You conclude that engineering and analysis is the most important criterion in the proposal and communications, the least. Chances are they have given each of these areas a weighting for scoring the proposals, but they are not telling you what that weighting is. Oftentimes they do, but they don't have to. Next comes a title: "Within Each Functional Area." So you have to deduce that the four listed items, "Engineering and Analysis," and so on, are the functional areas referred to in this title.

Your outline would probably look like this:

1.0 Engineering and analysis

     1.1   Approach to (engineering and analysis [E&A] function)

1.1.1   Perception of problem areas, and so forth

1.1.2   Management and supervision

1.1.3   Technical implementation (of E&A function)

1.1.3.1 System analysis (SOW para ____)

1.1.3.2 Systems Engineering (SOW para ____)

1.1.3.- [These four-digit paragraphs are where you address each of the SOW requirements in turn. This is where you *prove*:

"Feasibility of technical approach . . ."

"Degree of assurance that you can actually perform" etc., as stipulated in the evaluation criteria]

1.2 Personnel

1.2.1   Quantities and skills (just for E&A, remember)

1.2.2   Qualifications

1.3 Experience

1.3.1   Recruitment for overseas operations (for E&A functions)

1.3.2   Developmental engineering

2.0 Management and support

2.1   Approach

(And so on, as above, for management and support referring back to evaluation criteria)

It seems logical and straightforward enough until you get down to paragraph 3. "Experience," of the evaluation criteria, (para. 1.3 of the preceding outline). How, for instance, do you discuss "Recruitment for Overseas Operation" in the engineering and analysis section, or in the instrumentation operation and maintenance section? Or "Procurement of Complex Equipment" in these sections? It would seem these are strictly management and support functions to be addressed *in toto* there, and not elsewhere.

You have two choices: You organize the proposal just the way they say, as logically as you can, or you try to get clarification at the bidders conference by submitting a written question. Under paragraph 1.3 of the outline above, you tell about your company's experience in recruiting overseas for engineering and analysis people. You tell about how your company has procured complex equipment for performing engineering and analysis functions on other contracts, if you can possibly think of something that makes sense, and then you do the same for management and support instrumentation O&M, and communications, because that is what they asked for, and that is what they are going to grade you on.

The second kind of criteria—the standards—are the criteria that the Government evaluators use as a guide for evaluation when they read your proposal. Evaluation criteria are published in the RFP; standards are not. *It is up to you as proposal writer to deduce what these standards are, so that you can write your proposal in response to them.* If you fail to address any standard, you are guilty of a deficiency and marked down accordingly.

Standards are written statements of conditions necessary to achieve minimum acceptable performance. Standards are the measurement guides—the detailed instructions provided to the evaluators to help measure the adequacy of the proposal in meeting the requirements set out in the RFP. The evaluators will have a copy of the appropriate standards in front of them when they evaluate your proposal.

Likewise, standards will not address any requirement unless it is specified in the RFP (SOW, RFP instructions, criteria). The customer can't expect you to bid on specifications that they did not ask for in the RFP.

Standards must measure acceptability of the proposed solution. This is to say that any criterion that measures anything other than acceptability (that is, to achieve minimum acceptable performance) cannot be considered a standard. Conversely, a proposal that exceeds minimum acceptable performance cannot receive a higher grade than one that just meets acceptability. Presumably, if the customer wanted the higher performance they would have so stated. The message here is simply: "Don't think you will get a higher grade by proposing something nice to have and thus beat the competition. It may not do you any good. Likewise, it won't do you any harm either, provided it doesn't cost any more."

Standards should be prepared by the SSEB before issue of RFP. They *must* be prepared before receipt of proposal to avoid charges of fraud.

Again, failure to address a standard will constitute a deficiency. So you must analyze the RFP and deduce the Government's standards. You should be able to do this by taking each of the criteria in turn and identifying each of the SOW items that falls within each criterion, and determine what you must write to fulfill the criteria.

A standard is set out in four parts: area, item, factor, subfactor (Air Force terminology; other departments have similar terminology). Evaluation is made at the factor and subfactor level, the area and item elements being merely titles or headings for management purposes. As in the example following there may be several factors and subfactors within an item. The factor/subfactor descriptions will contain detailed descriptions or instructions for evaluating these criteria.

Example of Standard from Air Force Procurement

Item: Design/develop communication subsystem

Factor: Technical control element. Evaluated IAW factor description

Factor: Voice communication element

Subfactor: Automatic audio switch. Evaluated IAW subfactor description

Subfactor: Emergency manual service

Factor/subfactor descriptions are detailed evaluation criteria.

The following is an example of a NASA standard prepared for evaluation of the Key Personnel section of a proposal. This is what the evaluators have propped in front of them as they evaluate and score your proposal.

D. *Key Personnel*—150
1. This area should be evaluated considering the individual's qualifications, training, education, and similarity of past experience in comparison with the work required under this effort. Questions to be asked are:
   a. How recent has been his experience?
   b. Has the individual dealt with labor organizations?
   c. Is he currently employed by the offeror?

    d. Is salary proposed consistent with present earning?

    e. Is the rationale for proposing the man reasonable?

    f. Does the man meet the requirements as outlined in the RFP for the position?

    g. Scope of experience comparable to our effort?

2. Reference checks where possible will be made on key personnel. No specific weight will be given on these checks but they will be used to confirm or adjust the initial scores on the basis of the evaluator's judgment.

3. No specific weight will be given for availability, and no points will be added for key personnel being committed to the specific job. If, however, key personnel have not been contacted or committed to the job, a penalty will be assessed to the score for this individual. The penalty will be a judgmental factor made by the evaluator on the basis of criticality of the position and likelihood of a suitable substitute being available.

End of Example of NASA Standard

The third kind of criteria—those that no one sees: biases, prejudices, reputation, and the like—are the most difficult for the proposal writer to overcome. These "criteria" have been established before the RFP comes out—by past performance of your company, by personality conflicts or accommodations with the customer, and especially by the "up-front" marketeers with your company. These are the people that provide, to a great extent, the image that your company has with the key customer management and decision-making personnel. This subject is covered in detail in Chapter 1 of this book, but suffice it to say at this point that these people can, and do, exert a great influence, positive or negative, upon the success of your company in acquiring contracts. You, the proposal writer, can also create a favorable bias by preparing a neat, well-formatted, easy-to-read, straightforward, easy-to-evaluate proposal. Be aware that neatness is not just an esthetic labor. It could be the difference between winning or losing, because no one wants to read a messy, disorganized proposal. The evaluator will just take one look at it, give you an average score, and move on to the next proposal.

Summary: Proposal Evaluation Procedures

1. The establishment of evaluation criteria set out in RFP is an SSEB task, approved by the SSAC.
2. The SSEB also establishes the standards by which the Evaluation Team evaluates your proposal.
3. Standards are written statements of conditions necessary to achieve minimum acceptable performance.
4. Standards must:
   a. Not exceed specified minimum requirements;
   b. Not address nonspecified minimum requirements;
   c. Measure acceptability of proposed solution.
5. Standards must be prepared before receipt of proposals.
6. Proposals are evaluated against pre-established criteria and standards.
7. The SSEB establishes the competitive range.
8. The SSEB conducts orals (Q&A sessions) and establishes definitive terms.
9. The SSEB prepares the SSEB evaluation report.
10. The SSEB briefs the SSAC.

RATING TECHNIQUES

All instructions to evaluators are included in the source selection plan, including instructions for rating. For very large procurements, the SSEB is organized into various hierarchies to facilitate evaluation and surveillance of the evaluators. *Areas* may be in overall charge of managers (selected by the SSEB chairman). *Items* are assigned to "item captains"; *factors* and *subfactors* to "evaluators." If both narrative and numerical scoring is used, the evaluator usually provides the numerical rating, and the item captain writes the narrative. Color coding (in lieu of numerical scoring) is sometimes used, especially in the Air Force. This facilitates quick-look review by the SSAC or SSA where many proposals are evaluated for one procurement.

Once a proposal has been scored, no changes in scoring are permitted *except* to evaluate changes to the proposal, made as a request by the customer at the Q&A session (orals). In that case, a narrative justification should accompany any change in the numerical score.

Costs are never scored in Air Force procurements, but in other agencies, for example, the Navy and the Department of Commerce costs may be scored according to a pre-arranged formula, which may or may not be published in the RFP.

Figure 5-3 is one example of a rating sheet for numerical rating of proposals. There is no standard form for all proposals, the format usually being left up to the chairman of the SSEB. Note that each subfactor is given a percent weighting and that each sub-subfactor is provided a square for weighting within the factor.

Summary: Rating Techniques

1. Included in source selection plan
2. Flexibility
   a. Narrative
   b. Combination of numerical scoring and weights with narrative assessments
   c. Color coding with narrative assessments
3. Rate proposals as originally submitted
4. No rescoring or recoloring
5. Analyze changes narratively

CONTRACT CONSIDERATIONS

The model contract contained in many RFPs must be considered in writing your proposal. The purchasing and contract-

FIGURE 5-3 Rating Sheet

Date:	RATING
Item:_____	.0 Unacceptable
Spec	.1 Poor
	.2 Below Average/Marginal
Bidder:_____	.3 Average
Evaluator:_____	.4 Above Average
	.5 Excellent

FACTOR #3—MANAGEMENT	MAXIMUM	ACCEPTABLE	RATING	WEIGHTING	SCORE
A. PROJECT MANAGEMENT = 30%					
1. Degree of Managerial Support					
2. Number and Specialization of Personnel					
3. Qualification of Key Personnel					
B. MANAGEMENT APPROACH = 30%					
1. Project Schedule Controls					
2. Coordination of Activities					
3. Vendor Control and Liaison					
C. MANAGEMENT CAPABILITY = 40%					
1. Previous Related Experience					
2. In-House Capability					

ing officer (PCO) will compare proposals with the model contract to verify that there are no conflicts. Remember, your offer (proposal) coupled with the customer's acceptance constitutes a contract. If your proposal is inconsistent with the model contract, or fails to acknowledge substantive obligations contained therein (for example, the contract data requirements list (CDRL)), you are, in effect, taking exception to a portion of what the customer said he wants. If the above discrepancies are minor in nature you will be so advised during the orals and given a chance to amend your proposal accordingly. You may take exception to any portion of the model contract or the SOW, of course. But don't do it if you want the contract!

The PCO will also look at the price quoted to determine if the price is realistic in relation to the contract. If not realistic, your proposal will probably be peremptorily dropped from the zone of consideration, whether it be unrealistically high or unrealistically low.

DEFICIENCIES

A deficiency is defined as any part of an offer or proposal that fails to meet the customer's minimum requirements established in the solicitation.

Any one of the four conditions following may constitute a deficiency. If deficiencies found in a proposal are such as to constitute a fundamental flaw in concept, or otherwise are of such a magnitude as to be incapable of being corrected without major revision, the proposal will not fall within the zone of consideration (competitive range) and will therefore be eliminated from further consideration.

A deficiency is any part of a contractor's proposal that:

1. *Fails to meet* the minimum requirements represented by the standard; or
2. Proposes an approach that *poses unacceptable risk*; or
3. Is an *omission of fact* that makes it impossible to assess compliance with the standard for that requirement; or
4. Describes an approach taken by the offeror in the design of its system that *yields undesirable performance.*

Minor deficiencies may be corrected after the Q&A session. The greater the number of proposals submitted, however, the greater the chance of your proposal being eliminated for deficiencies.

Many RFPs will list major critical areas and stipulate that a failing score in *any one* of these major areas will disqualify your proposal from the zone of consideration, no matter how well you score on other areas.

I used to get very indignant when I first started writing proposals, when the term "deficiency" was used in debriefings or Q&A sessions. Until I realized, that is, that deficiency is a standard term used by proposal evaluators everywhere to identify portions of a proposal that fall within any of the four conditions set out above. It doesn't necessarily mean that your proposal is deficient, or that you are mentally deficient. I doubt if there has ever been a proposal that received a perfect score.

Since you, the proposal writer or manager, seldom get a chance to see the standards prepared by the procuring agency, I will include more real-world samples of standards at the end of this section for your study and future reference. You can readily see that it would be difficult to anticipate every single item that is listed in these standards, but you must make the effort. The effort involves study and interpretation of the RFP, thorough research and brainstorming of your technical approach, identification of critical areas, and constant vigilance to ensure that you have covered everything requested, directly or by implication, in the RFP.

Let us briefly consider each of the conditions.

1. *Failure to meet minimum requirements represented by the standard.* Suppose you propose a program manager who may be perfectly qualified for the job, but unfortunately he doesn't happen to have ever been in a position where it was necessary to deal with labor unions. If your evaluator is using the standards set out in a preceding section as an example of a NASA Standard for Key Personnel, then you would have a deficiency. The customer feels that experience in dealing with labor unions is of sufficient importance to be a factor in scoring your proposal (see paragraph 1b [page 136] of the sample standard). This would not be regarded as a serious deficiency, and certainly would not (in this case at least) disqualify you. But your score would be diminished to some extent.

2. *Proposes an approach that poses unacceptable risk.* This is not an attempt by the customer to discourage innovative approaches. What it does mean is that you have to do a thoroughly meticulous job of convincing the customer that your approach is feasible, that there are no uncertainties (for hardware systems), and that the design is feasible, that you have tested it, and that you have the documentation to prove it. If you cannot eliminate the uncertainties, then you had better propose a different approach. You can't expect the customer to buy a "pig in a poke."

3. *In an omission of fact that makes it impossible to assess compliance with the standard for that requirement.* This is the most pervasive of all deficiencies, and paradoxically it is the one most easily avoided. It simply means that you have to address *all* requirements set out in the RFP including all facts and data to enable the customer to fully evaluate your proposal. If you fail to address any particular of the RFP requirements, expressed or implied, you can be certain of getting yourself a deficiency. That is why I have been harping on this aspect of proposal preparation throughout this book.

4. *Describes an approach taken by the offeror in the design of its system that yields undesirable performance.* This is like item (2) above except that in that case the customer has uncertainties. In this case he knows it won't work, either because he has seen it tried before, or has proved to his satisfaction that it won't work. As a general rule, I would advise that you avoid any way-out unorthodox solutions to problems. The people who evaluate proposals are generally engineering or science-oriented or business-type people. Such people are usually conservative minded and have a show-me attitude. If they weren't that way, they wouldn't last long.

The remainder of this section consists of an excerpt of standards employed in evaluating a Government contract. It is included here to give you a better idea of the kind of criteria the customer may devise in order to make a thorough and detailed evaluation of your proposal. Remember, these standards are the detailed criteria prepared by the customer for the sole purpose of enabling the evaluator to score your proposal objectively. These standards are never revealed to the public; they are for the sole use of the customer—the procuring agency. They are devised to assist the evaluator in judging how well you responded to the evaluation criteria published in the RFP.

Do not conclude that the standards shown here will fit any situation. You must devise your own set of standards to fit each individual proposal effort. Your success in avoiding deficiencies then will depend on how close you came to matching the standards the customer devised, and of course, how well you fulfilled these standards in your proposal. You must consider the sample provided here as a guide and a starting point in developing your own standards.

EXAMPLES OF STANDARDS AND WEIGHTING FROM NASA PROCUREMENT

A. *TECHNICAL AND MANAGEMENT UNDERSTANDING—* 450
   1. *Operating Plan—*300
      a. Supervisory authority and responsibility
         (1) Does offeror set forth supervisory responsibility for each succeeding layer of organization?
         (2) Does each functional manager have the necessary authority to carry out his proposed assignments?
         (3) Are supervisory positions consistent with functions requiring monitoring?
         (4) Are supervisory positions described in the text consistent with organization chart?
      b. Managerial control devices
         (1) What techniques are proposed for each level of supervision to monitor succeeding levels of supervision?
         (2) How will parallel functional supervisors communicate with each other on matters of joint concern?

  c. Work flow and assignments
    (1) What procedures are set forth for how assignments will be made?
    (2) How are work priorities established?
    (3) Who establishes priorities?
    (4) What provisions are made for assuring a closed loop?
    (5) What techniques are proposed to determine assignment status?
    (6) How is status reported to NASA Director on a timely basis?
  d. Cost controls
    (1) What are proposed techniques for controlling expenditures made on equipment maintenance and supplies?
    (2) What does offeror propose as a method of assuring utilization of existing equipment?
  e. Training
    (1) What is proposed for initial training for new employees in each functional area?
    (2) What provisions are made for inservice training?
    (3) Is cross-training to be accomplished? How?
    (4) What is proposed for (2) and (3) above for supervisory personnel?
    (5) Does proposed depth of training meet Government needs?
  f. Does proposer cite any pivotal or crucial factors for a successful operation?
  g. Overtime
    (1) What are the proposed controls concerning overtime utilization?
    (2) At what levels of authority may overtime be approved?
    (3) At what levels of supervision will overtime not be allowed?
    (4) What is the proposed policy concerning the supervision of those assigned to unique or one-of-a-kind tasks?

    h. Emergency planning
      (1) Does proposer present a plan for major fire or disaster?
      (2) Does plan provide for cross-utilization of manpower?
      (3) Does plan provide a means of call for assistance beyond company capabilities?
    i. Phase-in
      (1) Does proposer describe a phase-in plan that will provide for optimum operations during transition period?
      (2) What other manpower resources will be made available from within company?
      (3) Is a plan provided for recruitment of new personnel? Is it adequate?
      (4) Is sufficient time provided for relocation of personnel to accomplish transition?
      (5) What percentage of existing work force is proposer planning to recruit?
    j. Utilization of personnel
      (1) What plans are offered concerning interface between off-site and on-site personnel?
      (2) What plans are offered to permit flexibility to meet shifting or peak work loads?
2. *Organization and Management*—150
    a. Organization
      (1) Are the functions assigned to each level of organization clearly identified?
      (2) Are the functions logically grouped? Is there unnecessary duplication of functions?
      (3) Will the proposed groupings create problems in accomplishing the work?
      (4) Are the proposed lines of communications within the project organization simple and direct?
      (5) What is the relationship of the project organization to company or home office organization?
      (6) Is the plan for subcontract management effective?

b. Management
   (1) Does the project manager have the necessary authority to do the job?
   (2) Do the supervisors of major functions have the requisite authority?
   (3) What support is contemplated from the front office? Too much? Too little?
   (4) What management techniques will be used at the project-manager level to ensure success of this effort.
   (5) Are the interface relationships between the government and contractor organizations on-site adequately covered? Are these relationships workable?

End of Example of Standards and Weighting

CHAPTER SIX

SOME FATHERLY ADVICE FOR PROPOSAL MANAGERS

Managing a proposal is essentially no different from managing any other momentous project. The same principles of good management apply in either case. As a good manager you will, in managing the project:

- Establish the extent of your authority and responsibility;
- Prepare yourself with in-depth study of the problem;
- Make detailed plans for carrying out your objectives;
- Break down the task into logical, manageable segments;
- Make a careful selection of qualified and reliable key personnel to assist you;
- Assign responsibility and delegate authority to key personnel to accomplish appropriate segments of the task;
- Establish a *modus operandi* for accomplishing the task;
- Establish interfaces required for coordination;
- Establish a control and feedback system to ensure timeliness, quality, and early identification of problem areas.

Now let us examine each of these precepts as applied to proposal managing.

Establish the Extent of Your Authority and Responsibility

Let's face a few unpleasant facts of life right at the outset. Managing a proposal is not a job for the faint of heart, or, I might add, the thin of skin. It is hard, demanding work and usually a thankless job (though it shouldn't be). Confucius (or maybe it was Chairman Mao) once said, "Success has a thousand fathers; failure is an orphan." Nowhere in the business world does this aphorism apply so accurately as in proposal writing. If you succeed in winning the contract, people start coming out of the woodwork to claim the credit and you will be amazed at how many different factors accounted for the win. But if you lose, no matter how the contract may have been wired for someone else, or how badly your corporate management screwed it up, the fingers will all point at you, the proposal manager.

Inasmuch as you are destined to take the blame for losing, the least you can do to protect yourself is to *demand* that you have authority to control the proposal process. To start with, you should have a strong voice in making the Bid/No Bid decision. Apply the criteria listed in Chapter 1 in making your own Bid/No Bid decision. If now you agree that it is a viable opportunity, you must demand that management provide you the resources to accomplish your goal, specifically:

1. *An adequate budget and a completely free hand in managing the proposal.* You, the proposal manager, must make your own budget estimate, based on the requirements set out in Chapter 1, and any other special requirements you can anticipate. Be objective and realistic. Almost everyone has a tendency to get too euphoric at this stage, especially if you have a gut feeling you want to go after the contract. Be candid with management in delineating all cost factors involved in arriving at your budget estimate. It's best to get a clear understanding of this matter at the outset. Otherwise you will end up spending a lot of your precious time begging for more money as the proposal effort progresses, and they will never forgive you for "conning" them into this venture under false pretenses.

Once you have got a firm commitment from management on budget, you must ensure that they understand that you must have a free hand in how, when, and where to spend the money. Of course, you must keep them informed, even on a daily basis if they desire, but the spending decisions must be yours alone, without any meddling or sideline coaching.

2. *A firm understanding with management on just who you can have on your Proposal Team.* (Get this, "I can't spare Joe Zoaks" stuff from line managers out of the way in the very beginning.) Presumably your company has made a Bid decision, based on a viable business opportunity with all the benefits that inure from a contract award. Well, you can't have something for nothing. Many people must share in the sacrifices to make this opportunity a reality. Don't let them put you in the position of being the only one making a sacrifice—or jeopardizing the success of your mission by having to accept incompetent, second-rate people to assist you. Field managers will, with almost kneejerk consistency, balk at letting you use any of their key personnel to help on a proposal, claiming that those people's absence would ruin their operation. Balderdash! The reality is that they simply don't want to be *inconvenienced.* Ask them what they would do if that particular person got sick and went to the hospital for a month or two. Would his operation fall apart? If so, then he is a lousy manager. Any good manager prepares for such contingencies by grooming potential successors to all his key personnel.

3. *The authority to propose whomever you want (within reason, of course) as key personnel on the contract.* The same reasoning as in the previous paragraph applies to whom you will propose as key personnel. You can expect some managers to scream bloody murder if you do anything to disturb their comfortable *status quo.* Let them holler. You already have secured the backing of top management—haven't you? Obviously, these selections and decisions must be exercised with discretion, restraint, and good common sense judgment. You can't go on raiding parties—spread the burden around.

Incidentally, you generally will not be using all the key personnel you propose if you win the contract. You certainly must produce the proposed program manager, as well as those proposed for certain company-sensitive positions like finance and administration director, for instance. But you will actually be picking up the incumbent personnel for many key positions, depending on the circumstances. This is the program manager's decision after award of the contract.

4. *The authority to organize the proposal team and to draw on company support* (typists, word processors, reproduction, editors, and so forth). Nine times out of ten, the company management will underestimate the magnitude of the logistical and administrative support required to prepare a proposal. I once managed a large proposal where I ran into this problem. I'm not recommending this as an ideal way to handle the situation, but this is the way it went. I made an impassioned plea to the executive vice president for dedicated word processors and assignment of a proposal coordinator (a key person in any large proposal, as described later in this chapter). He reluctantly offered me *one* dedicated word processor. All he could spare from his force of some 400 people for a proposal coordinator was—get this—the kid who took care of the coffee shop, the vending machines, and the Xerox® machine. So, tired of making impassioned pleas, I decided, "OK, *you* are ultimately going to be held accountable for this proposal, so you can learn the hard way." Of course this proposal progressed into the predictable chaotic situation, and we eventually had to commit *three* word processors with operators working *two* shifts and an engineer/administrator who took over as proposal coordinator working some 14 to 18 hours a day.

You see how proposal costs can go out of sight simply because of poor planning and organization? Even worse, you are bound to end up with a sloppy proposal and some very disgusted working troops.

By the way, I hate to tell you this, but we won the contract in spite of such incompetence, and the executive vice president has been basking in glory ever since. I bade a not-so-fond farewell to this outfit soon afterward, but I would surmise that the kid who ran the coffee shop is now director of business planning or something like that.

The foregoing experience reminds me of a bit of persiflage making the rounds in this business. If you've heard it before, forgive me and skip on.

The Ten Phases of Proposal Preparation

1. Euphoria
2. Exhilaration

3. Uncertainty

4. Confusion

5. Disillusionment

6. Panic

7. Search for the guilty

8. Punishment of the innocent

9. Banishment of those who toiled

10. Rewards for the uninvolved

Please don't take this too seriously, but it's good to remember that it's hardball they are playing out there.

5. *The authority to replace inept members of the proposal team.* Several years ago I worked on a proposal that went along very smoothly except for one thing. We had one key member of the team who was totally inept at writing. It was clear from the outset that he would never be able to hack it. Yet he was never removed from the team, because, inexplicably, he was the fair haired boy of the *company president.* So, in the end, all the rest of us had to pitch in and rewrite all of his stuff, whether we knew anything about the subject matter or not. I remember I had to find a book on sewage disposal (which I know absolutely nothing about) and write a learned treatise on this subject as applied to this particular contract the day before the proposal was due. What fun!

6. *Authority to determine the proposal strategy and basic approaches.* I'm talking here about major proposal decisions: contract organization and staffing, degree of program autonomy, corporate interfaces, where the work will be performed (if there is a choice), budgetary controls on the contract, and where and under what circumstances the proposal will be prepared. These determinations are not made in a vacuum. They must be made as a coordinated effort with full discussion between you and corporate management. You want to be sure you are included in these discussions lest you be saddled with some hip-shooting decisions that you cannot live with.

7. *A strong voice in plotting cost strategy.* Same comment as above. You may not be an accountant, but you know more than anyone else about the criteria for costing this contract. So, establish open lines of communication with the person charged with costing this contract, early on, *and* with the finance director, controller, or whatever, and insist on participating in any conferences on cost strategy. After all, you are responsible for the success of the *whole proposal*, so don't let them ram any unacceptable costing decisions down your throat. I'm referring to G&A, fee, award fee structure, labor rates, other direct costs, and so on. Believe it or not, I have managed proposals during which I was never consulted on any of these things—and most of them were losers.

8. *Authority to make all arrangements for use of consultants.* There will be more advice on consultants later, but the point I want to make here is that you, the proposal manager, must control their use, and for that matter, make the decision on whether you need them in the first place. My first experience with consultants has perhaps given me a bias against them in general. I was managing a large proposal effort, reporting to a company vice president, who apparently knew a lot of consultants but not anything at all about preparing proposals. I was virtually inundated with unwanted and unnecessary consultants. I was daily besieged by more conflicting advice from every side than I could even sort out. None of them wanted to write anything—just talk, talk, talk. None of them came prepared, submitted any documentary material, made any prepared formal presentations, or even provided any insight into interpreting the RFP or the nature of technical problems. The whole proposal was in imminent danger of ending in total confusion until I succeeded in running them all off. You must make the decision whether you need them at all, and then, whom you need, for how long, and when.

9. *Your constitutional rights* . . . To freedom from harassment by management while the proposal is being prepared. This means keeping the briefings and status reports and management meetings to a minimum; keeping unwanted consultants, review teams, and other management spies off your back; and scheduling management review at the time of *your* choosing.

The biggest menace the aspiring proposal manager must face is the well-meaning, but inept and insecure company management team that do not know how to do the job themselves but won't let you alone to do the job yourself. They keep calling status meetings two or three times a week, so that you can explain why you are behind schedule because they keep tying up your time with these stupid meetings. They start bugging you to produce something for them to "review" before you even have a chance to complete your detailed outlines. They send spies from the corporate office, disguised as "consultants," their mission being to report back on whether you are doing your job right. They send out so-called Red Teams who never analyzed the RFP. (Red Teams or Tiger Teams are usually hand-picked representatives of the company management whose function is to review your proposal and often to play "devil's advocate." I have a few choice words on this subject in Chapter 7.) Their mission is to tell you how you should have written the proposal, based upon some kind of mystic insight, I suppose. They change basic strategy on you midway through the proposal. Once I even had a proposal effort peremptorily moved—lock, stock and barrel—clear out of the continental United States, 4000 miles away, right in the middle of our schedule!

I had the unpleasant experience once of trying to manage a proposal for recompetition of an existing contract. The program manager couldn't write a coherent Christmas card to his grandmother, much less contribute anything to either writing or intelligently reviewing the proposal. But he just had to keep his hand in. He would tell me, "I just have an *intuitive* feeling this management plan (instrumentation plan, maintenance plan, it didn't matter) is not very good." Then he would call in consultants one after another until he found one that corroborated his fabled intuition. I think that proposal was rewritten seven times. Funny thing, though. What we finally ended up with was the version we started out with in the first place.

The thing to do is set up a schedule for the proposal effort as soon as possible after the RFP comes out and disseminate it to all the working troops *including management*. Make sure the schedule provides for status meetings, periodic progress reports, briefings, and a formal management review. Then make it abundantly clear to all concerned (especially management) that this schedule does not have room for interruptions, perturbations, or tinkering. Failure to adhere to the schedule will jeopardize the entire effort, so *don't mess with it.*

If you cannot get a clear understanding on all these items before you start, you had better start updating your résumé, because otherwise you cannot possibly do the job adequately, and you will end up being a sacrificial goat. Management has chosen you to manage the proposal and you must demand the means with which to do the job.

Now, while we are on the subject of authority, a few more rules are in order.

a. If your company is the incumbent, the incumbent program manager should be the proposal manager.

b. If your company is not the incumbent, the *proposed* program manager should be the proposal manager.

These two rules apply whether these proposal managers know anything about managing proposals or not. The reason for this is discipline. The proposal team must be a disciplined organization in order to put in the long hours of coordinated effort required of it. In either (a) or (b) above, this discipline can be achieved because in (a) you are using an organization and chain of command already in being (the proposal manager chooses the proposal team from among the key personnel on the contract.) In (b) the same result is achieved if you have a chain of command that *will* be in being if awarded the contract. (Again, the proposal manager should choose for the proposal team the key personnel who will be assigned to the contract.) An example is shown in Figure 6-1.

What if the proposal manager, then, does not know anything about writing a proposal? That is where good corporate management comes in. Ideally, every well-managed company should have at least one individual on its staff who is expert in proposal writing and managing. He doesn't have to be strong technically, although an engineering background is desirable.

FIGURE 6-1 Proposal Team Chain of Command

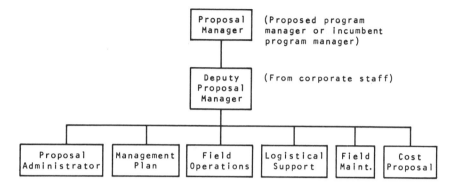

But he *must be expert* in proposal preparation. Then this person can manage the proposal himself through the titular proposal manager (providing the titular proposal manager has the good sense to stay out of the writer's way so he can apply this special expertise without hindrance).

The deputy proposal manager will in this case be doing most of the work in close conjunction, however, with the proposal manager. This does not mean the proposal manager can remain aloof from the day-to-day problems of the proposal. He is ultimately responsible for its success; he must make the basic decisions on technical and management approach, because he is the one who is going to have to live with them. Also, don't forget, the *proposal manager* is the one who will have to appear at the orals and get a going-over from the prospective customer, so he had better become intimately familiar with the proposal as it progresses.

The deputy proposal manager can be of invaluable assistance in preparing the proposal plan, writing the executive summary, assisting the proposal coordinator in arranging support requirements (typing, security, reproduction), and most especially in providing guidance to the team in proposal preparation, critique, and correction of the first drafts. In fact, this person can contribute so much that management should consider providing someone like this *in all cases*—even if the proposal manager is an expert writer. This would provide a counterpoint to proposal managers who may pay too much attention to technical details and thus get bogged down in relatively unimportant matters, losing sight of the big picture.

A word of caution here in choosing key personnel to propose for managing the target contract. You must consider the political, psychological, and other factors involved. By way of illustration, I once managed a proposal for operation and maintenance of a large, sophisticated meteorological program managed by the Department of Commerce. The operation involved taking weather data from geo-stationary satellites and distributing this data nationwide. No weather forecasting—just operation and maintenance of all this hardware/software. I thought it would be a good idea to propose someone I knew for deputy program manager. He had a degree in physics from the University of California (Berkeley) and several years of experience in a wide variety of hardware/software systems. Only trouble was he was also a meteorologist. No way. They turned thumbs down on him and I suspect the only reason is, they didn't want some contractor person around who might possibly know more about meteorology than they. So we lost and I learned a costly lesson. Listen, you must win the contract first; then you can work out the pragmatic details of how best to perform it.

DO YOUR HOMEWORK

You, the proposal manager, are the expert, the decision maker, the leader. You cannot be these things without doing your homework. You *must* get a head start on everyone else so you can become completely familiar with the background, the requirements, the politics—everything that is known about the contract before the RFP comes out. And when that happens you must go through the whole RFP, marking everything that needs to be addressed and then go over the important parts again and again until you thoroughly understand it, because you are going to be asked a zillion questions, both by your team and also by management. In the initial stages you will have to make many crucial decisions that cannot wait, and they must be made intelligently.

The first thing you should do when appointed proposal manager is to get together with the individual in marketing (or program development) who has been tracking this opportunity and pick his brains, read his file, and go with him, if possible, to the site for a preliminary look-around and a talk with the contracting officer. If you are lucky, you might find out some more information from him (the contracting officer), but at least it will give you an opportunity to size him up and perhaps impress him with your interest and sincerity.

Next, start gathering all references, old proposals, correspondence, organization charts, résumés, reports—anything that can have a bearing on the proposal at this time. Familiarize yourself with what you have gathered, because you want to assign various parts of this material to your proposal team later, to prepare them for their writing assignments.

Finally, get in touch with other people either in-house or in other companies who have worked there before, managed similar contracts, have personal contacts, or have knowledge of the operation (such as vendors, subcontractors, or retired Government employees, or especially, defectors from the incumbent contractor). If it is a particularly sophisticated or state-of-the-art type of program, it may be necessary to get out some textbooks and bone up on the technical aspects of the program.

From the very beginning you must prepare yourself to organize your proposal team, identify the key personnel on the contract, and make tentative selections of which résumés to use in filling these positions. You must be prepared to make out a proposal schedule, a proposal plan, identify proposal themes, sketch out the executive summary, and chair meetings with your proposal team and with your corporate management. You have to provide guidance and direction to your team leaders and your proposal coordinator, and establish interfaces with the costing people and with those responsible for providing logistical and administrative support. Nobody can do it for you. Obviously, all these tasks require in-depth study, analysis, and preparation. Don't try to wing it or play it by ear. If you do, both your proposal team and your management will soon lose confidence in you, and that could be fatal.

To repeat, you must in all ways be the expert, and you cannot fulfill this role without doing your homework.

MAKE DETAILED PLANS FOR CARRYING OUT YOUR OBJECTIVES

The "detailed plans" will be embodied in a proposal plan. This is a formal document setting out all the information and instructions that anyone needs to know about the proposal effort—and don't try to write the proposal without it. You should start formulating your plan as soon as possible after you have been appointed proposal manager. You should plan on presenting it to the proposal team as soon as it is organized—at your first meeting. As more information is obtained, some deferred decisions resolved, further coordination achieved, you incorporate these changes or additions as addenda to the original document. The proposal plan is the basic document that guides all those involved in the proposal effort from inception to conclusion so that everybody is singing from the same sheet of music. It is the document that contains all official instructions, disseminates all significant information, and coordinates all support requirements for the proposal.

As a minimum, the proposal plan should include the following sections:

Basic Assumptions: All information of importance to the proposal team that you may not have now but is required as a working concept, for example, date of RFP release, cost-plus or fixed-price contract, turnaround time for proposal submission, location of proposal effort.

Organization of Proposal Team: An organization chart showing names of all key personnel and their assigned responsibility. Everyone involved must know who reports to whom and for what. Also, the authority vested in key personnel, for example, "Key personnel will have authority derived from top management to request information or inputs from whoever has such information, wherever located in the company."

Modus Operandi Prior to RFP Release: General plan for preliminary preparations for proposal: schedule of briefings, classes in proposal writing, availability of reference materials, assignment of preliminary tasks, gathering of résumés, reference material, marketing intelligence, customer contacts, indepth studies . . .

Proposal Instructions: Before the RFP comes out, a brief statement of the concept for proposal preparation based on available information such as the old RFP, the current contract, and so forth. After the RFP comes out, all instructions may be set in concrete. A general outline will be prepared for the entire proposal. Special instructions implementing RFP instructions will be issued. Style guide, format, and other such details will be published as appendices.

Security Procedures: Control, safekeeping, destruction, and handling of proposal inputs; security of working space, control of personnel, and the like. (Proposal material should generally be handled like secret material.)

Administrative Procedures: Arrangments for typing, editing, use of word processors, printing, cost controls, routing of inputs, review and correction, accounting, drafting, art work, office supplies, travel, TDY, reports.

Schedules: You *must* provide a milestone chart for proposal preparation in your first edition of the Proposal Plan. Base it on an assumed date for proposal submission. Or just call RFP release date D-Day; the time intervals are the important thing. This gives everyone a finite period of time for planning purposes for completing various phases of the proposal. Upon study and reflection it may become apparent that the schedule may have to be revised in some respects, but this gives everyone a frame of reference to start from. When the RFP does come out, *then* you should be able to put out a firm schedule (as an addendum) and demand that everyone strictly adhere to it. It is never too early to remind everyone that all must live within a strict schedule (and keep reminding them). The tendency every time is to start out too slowly and then end up in a panic.

The following pages contain a sample of a proposal plan for your reference. Of course, you may use any format or content that suits your needs. This sample is included here to show you the things a proposal plan should have and a suggested format. The sample milestone chart contained therein (as Enclosure 1) is likewise provided as a guide for future reference.

EXAMPLE OF PROPOSAL PLAN

ABC Proposal Plan

1.0 *Basic Assumptions*

- Expected RFP relase date: o/a _____.
- Proposal due date: o/a _____.
- Type of contract: CPAF (CPFF, FFP, or?) _____.
- Location of proposal preparation: _____.
- Location of Work: _____.
- Expected Competition: _____
- Value of the contract: about _____ covering the full five years of a 3+2 contract.
- The RFP will be essentially the same as the old RFP issued in _____.
- There will be a bidder's conference in _____ in _____. (month, year)
- The technical proposal will be limited to _____ pages.
- The contract will be managed as a separate cost center.
- There will be no teaming arrangements, only minor sub-contracting.
- There will be a _____ day phase-in.
- A thorough phase-in plan will be required.

2.0 *Organization of the Proposal Team*

- The proposal manager is Mr/Mrs/Ms _____.
- Deputy proposal manager is _____.

(See figure 6-2 for example.)

FIGURE 6-2 Example of Organization of Proposal Team

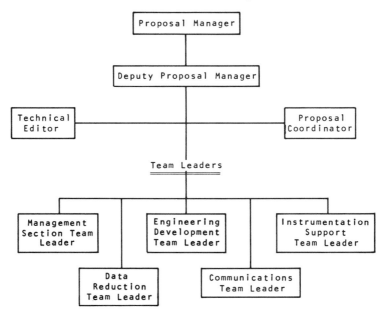

Team leaders will be responsible for all proposal inputs within *their* respective assigned areas. For example, the instrumentation support team leader will be responsible for everything listed in Section 3.0 of the SOW: telemetry, radar, photo instrumentation, and so on. This will include guidance to the various contributors, review of their inputs, dissemination of information pertaining to the proposal effort, ensuring that deadlines are met, and that all required interface with proposal management is accomplished.

The proposal coordinator will be responsible for all administrative and logistic support to enable the proposal team to function. This will include arrangements for equipment and supplies, drafting and typing support, security of proposal material, work flow control, assembling of final draft, printing, binding, and distribution. He or she will be responsible for maintenance of the master volume of both the first draft and the final draft.

3.0 *Organization of the Proposal*
The proposal will be organized into two volumes as follows: (Example)

Volume I: Technical Proposal
            Section 1. Management
            Section 2. Engineering/analysis
            Section 3. Instrumentation support
            Section 4. Data reduction
            Section 5. Communications
            Section 6. Ordnance

Volume II: Cost proposal

4.0 *Proposal Preparation Procedures*
(Example)

    4.1 *Page allocation:* Total 300 pages. Tentative allocation of pages is as follows:
            Section 1. Management, 65 pages
            Section 2. Engineering/analysis, 43 pages
            Section 3. Instrumentation Support, 101 pages
            Section 4. Data reduction, 20 pages
            Section 5. Communications, 60 pages
            Section 6. Ordnance, 11 pages

    4.2 *Format* (Example)

4.2.1 *A general outline* will be issued to all proposal team members within five days after RFP release. Each team leader will ensure that detailed outlines be prepared *in conformance* to the general outline and submitted on time in accordance with the milestone chart to be included herein.

4.2.2 Each major section, e.g., SOW 1, 2, etc., will begin with a *matrix of company-related experience pertinent to that sec-*

tion. Detailed related experience will then be presented in brief narrative form as the introduction to each of the respective subsections: e.g., Telemetry, Radar, etc.

4.2.3   In general, follow the same sequence in proposal contributions as appears in the SOW.

4.2.4   Each major section, i.e., SOW 1.0, 2.0, etc., will be separated by a tab divider. This tab divider will be imprinted with the overall organization chart for the contract, with the organization units *responsible for performance of those respective functions highlighted* to help the reader identify which organizational elements are responsible for performance of that particular function of the contract.

4.2.5   Hierarchy. Use the following hierarchy in use of job titles:

Program manager
(Department) managers
(Section) supervisors
Unit leadman
Senior technicians
Technicians A, B, C

See organizational chart on wall of conference room. For uniformity, use above hierarchy throughout. *Do not* try to use Wage Determination titles. This translation will be made in an appropriate place in the proposal.

4.3   *Schedules*
See enclosure 1 (Figure 6-3).

4.4   *Résumés* (Example)

Team Leaders will be responsible for providing résumés of *key* personnel within their area of responsibility. Résumés should be limited to two *pages.* For résumé format, see Enclosure 2. Team leaders in coordination with proposal manager will also be responsible for preparation of job qualifications in accordance with RFP requirements.

4.5   *Proposal writing instructions*
See Enclosure 3.

4.6   All questions on editorial convention and detailed format are to be referred to _____, (editor)

FIGURE 6-3 Enclosure 1: Milestone Chart

EVENT	D-20	D-Day	D+30	D+60
Make Bid Decision	(▽)*			
Designate Proposal Manager	(▽)*			
Research and Analysis	(▽)*	▽		
Visit Customer Facility	(▽)*			
Identify Key Personnel**	▽			
Prepare Proposal Plan	▽			
Kick-off Meeting, Proposal	▽			
Preparation Boiler Plate	▽	▽		
RFP Release		▽		
Reproduce and Distribute		▽		
Issue General Outline		▽		
Update Proposal Plan		▽		
Issue Proposal Packages		▽		
Meeting of Proposal Team		▽		
Bidder's Conference/Tour		▽ ▽		
Preparation Contract Organization		▽ ▽		
Preparation Contract Staffing		▽ ▽		
Preparation First Cut (Team Leaders)		▽ ▽		
Submit First Draft to Proposal Manager			▽	
Proposal Management Review			▽	
Corporate Management Review			▽	
Proposal Team Meeting			▽	
Status Review-Corporate Management			▽ ▽	
Correction/Revision Final Draft			▽ ▽	
Final Review-Proposal Management/Corporation Management				▽
Final Edit				▽ ▽
Printing and Binding				▽ ▽
Deliver Proposal				▽

*These actions are taken well in advance of RFP release, depending on the circumstances.
**Both for the contract and for the proposal team.

who is to be in charge of all editing matters concerned with this proposal.

4.7 Only two drafts are planned for this proposal. The first to be prepared *only* after exhaustive research and consultation with proposal management, after a preliminary detailed outline is prepared. This draft will be carefully reviewed, evaluated, corrected and returned to the writer with recommendations for resubmitting as a final draft. The final draft will be subjected to both proposal management review and corporate management review to be conducted concurrently. The next step will be the professional editing and preparation for publication as soon as all final corrections have been made. See Enclosure 4 for flow chart on proposal preparation.

4.8 Reference Material

Each proposal contributor will be provided a proposal package containing pertinent portions of the RFP, special instructions, and anything else deemed necessary to prepare the respective portions of the proposal. A library of additional information (old proposals, background material, reference material, studies, data, and complete copies of the RFP) will be maintained at the proposal management office. All personnel are urged to make full use of this material.

4.9 Addenda

This is to be considered an open-ended plan. As changes occur and new information is acquired, addenda will be incorporated in this plan, so that *all* instructions throughout the entire proposal cycle will be integrated into just *one* document, available to the entire proposal team.

End of Sample Proposal Plan

BREAK DOWN THE TASK INTO MANAGEABLE SEGMENTS

Almost without exception, the inexperienced or untrained proposal manager will do one of two things, both of which lead to big trouble. The usual case is trying to be a one-man band. Playing manager, editor, administrator, rewrite person, reviewer

—everybody's expert and making everybody's decisions. This is the syndrome of a conscientious, well-meaning person who doesn't understand the magnitude of the job. The other type, the poor soul who never got organized to begin with, is simply caught up in the inexorable swirl of events that engulf him, and he can only make feeble attempts to put out the fires as they break out and try to react to the relentless series of calamities as they inevitably occur. Both types are more to be pitied than scorned, because their fate is, at worst, to end up in the "funny-house" or, at least, to be relegated to the unemployment line in ignominious disgrace.

The message here is: Don't try to do it all yourself. It is impossible. You are supposed to manage. That means: plan, organize, delegate, direct, supervise, review, and correct. Do not try to write anything except the executive summary. Do not try to rewrite someone else's garbage. Don't be bogged down in administrative details (arranging for typists working over-time, ordering office supplies, that sort of nonsense). And above all, *do not try to play editor.* You do not have time for that.

Now, what are manageable segments? Suppose you have a proposal with a management plan, a technical proposal and an executive summary. The technical proposal may be broken down into three major sections, say sensor systems, communications, and systems engineering. So you must select key personnel accordingly. The management plan may include safety, security, personnel administration, résumés, subcontract plan, and procurement. But the only people *you* select are the ones responsible for the management plan and the one responsible for the technical proposal. They, in turn, will be responsible for selecting the individual in charge of the sensor systems section and the personnel administration section, and so on, with your approval, of course. The idea is to establish a working chain of command at once. You do have veto power over their selections and you must use it intelligently but freely. You do not want any deadbeats on the team who can't write, and in the final analysis it is up to you to see that you don't.

Now you have one more *very* key person to select, the person in charge of all administrative details, the proposal coordinator. Pick this one with care, because he or she can relieve you of a great deal of agony and keep you from becoming hopelessly swamped in a morass of minutiae from which you will never be able to extricate yourself in time to do what you are there to do—manage a proposal.

This person should be an experienced administrator capable of handling all types of people, devoted to detail, patient, persevering but decisive. Turn over all the administrative details to him: the typing, word processing, security procedures, control of paper flow (which can be a real headache in itself), editing, drafting and illustration, reproduction, printing, binding, correspondence, cost accounting, timekeeping, personnel, transfers, and I could go on *ad infinitum*. Tell this person to make all the decisions himself and delegate as he sees fit—just keep you informed on what he is doing.

Maybe some of the foregoing sounds like pretty elementary and obvious stuff to experienced managers. But I have time and again watched in growing apprehension and utter amazement as experienced managers tried to arrogate all these functions to themselves with predictably calamitous results. It isn't so much that they are performing many functions that can be performed by clerks. The problem is that they are perforce neglecting the job that everyone is depending upon them to do—manage the proposal.

CHOOSE YOUR TEAM MEMBERS WITH GREAT CARE

Joe Zoaks is a brilliant engineer. He can solve problems that other engineers didn't even know were there. He was born with a silver slide rule in his hand. But his boss says he doesn't like to write reports or document his work.

Clyde Fox isn't considered much of an engineer. He doesn't like detail, hates involved mathematical problem solving, just gets by technically. But his boss says he writes good reports, is very articulate, and seems to have a knack for expressing himself well, even about subjects in which he has little expertise.

Which one do you pick as team leader for the "technical approach" section of your proposal? Answer: Clyde Fox, without a moment's hesitation. Clyde can pick Joe's brains for what he needs to know about the technical aspects, but there is no way that you will ever teach Joe how to be a writer. Even if Joe has available some good writers to help him, somebody has to be able to differentiate between an articulate, well-written section and a cut-and-paste collection of garbage. Somebody has to know enough about writing to be able to provide guidance to proposal writers and to select good writers for his team.

Don't ever forget that writing is the name of the game in preparing proposals. It is the *sine qua non* of the proposal game. No matter how brilliant your ideas, they will go for naught if not expressed clearly and in an organized fashion. A proposal is the *written* response to the *written* word of an RFP. Nothing but the written word counts. So do not allow anybody with a writing assignment on your team unless he has a demonstrated ability to write. No matter that Joe Zoaks is a brilliant engineer, no matter that Charlie Shmo is a nice guy and wants to help, no matter that Ruby Klutz is the daughter of the contracting officer. Use them if you want, but not for writing *anything*, because they will be a continual drag on the whole team, and the harm they will do will be compounded as time goes on. As Harry Truman might have said, "If you can't write, stay out of the kitchen!"

It may seem that I have belabored the point a bit, but if so, it is with good reason. Too many times I have seen a proposal effort end up in a horrendous panic simply because at the last minute the proposal manager had to face up to the fact that Alvin Blurt couldn't write a postcard home to his mother without messing it up, and then everyone had to drop what he was doing and work all night to rewrite all of Alvin's garbage—under those circumstances, with deplorable results.

This is a good place to make another point—proposal manager, do not try to save Alvin's scalp by rewriting his junk yourself. To do so is to invite your own destruction for two reasons: (1) you will soon get bogged down in details at the expense of your main mission—managing, (2) even if you had the genius of Shakespeare, everyone wants to be a critic and soon you will have a bunch of self-appointed critics critiquing *your* work and ignoring your critiques of their work. Incidentally, managing a proposal is no place to practice experiments in democracy. There can ultimately be only one manager, only one person who sets the standards, and only one person who makes the day-to-day decisions. *You* are ultimately responsible and, if everyone marches to different drummers, you will all go down together. I might mention a third practical consideration: as soon as the rest of the proposal team finds out you are willing to bail them out when they turn in garbage, everyone will start to turn in garbage, knowing you will rewrite and rethink it for them. Do not let this scenario begin, because it will lead inevitably to a sad and tragic ending.

Before leaving this subject I must unburden myself of a few observations on the subject of who is or is not qualified to *write* contributions to a proposal. Many people will plead incompetence when called upon to write *anything*. If they are nonprofessional people, paid on an hourly basis, hired to perform operational functions not involving any writing skills, perhaps they have a point. But if they are professional people— degreed engineers, administrators, executives, supervisors, managers—enjoying all the perquisites of management people, what is *their* excuse? Presumably, if they have a degree in anything, they should know how to write at least a few coherent paragraphs in response to a Statement of Work. Surely that capability must be a qualification for graduation from any accredited college. If they reached their management position without a degree, surely they must be as smart as those who have one. The plain fact is that in most cases, people do not want to be dragged into a proposal effort because (a) it is a thankless job, (b) if it loses, they will be blamed (never mind that management should know the chances of winning are, at best, about 1:4 on average), (c) inevitably they will have to work much harder than on their eight-to-five job, and most likely longer hours, and (d) if they do a good job, they will be called upon to make these sacrifices over and over again because they are good. Well given the above, can you blame them? And sadly, (a), (b) and (d) are solely the result of poor management.

The judicious selection of competent writers for your proposal team is an area that is too frequently ignored. No less important to the successful and prudent proposal manager, however, is the selection of a top-notch experienced administrator (a proposal coordinator) to take all of the administrative details off your shoulders and leave you free to do what you are supposed to do—manage. This subject was discussed two paragraphs ago, and the only thing I would add here is: (a) Do make sure you pick the right man (or woman) in the beginning, because *this* choice is almost irreversible, and (b) take positive steps to ensure that he coordinates with you and keeps you informed. Even the best administrators sometimes get drunk with power or go off on their own, doing their own thing. Ask Richard Nixon for more details on this point.

ASSIGN RESPONSIBILITY AND DELEGATE AUTHORITY

You will note that I said *assign* responsibility and delegate *authority*. You cannot delegate responsibility. *You* are irrevocably the only one responsible. But, you can assign responsibility and you can delegate the authority necessary to fulfill that responsibility. Responsibility is one side of the coin; delegation of commensurate authority is the other. I made a big point of ensuring that you get both sides of the coin in the opening section of this chapter. See that your subordinates get the same fair shake.

Your team leaders and proposal coordinator may often have to cut across the chain of command in order to get the necessary information and support. Oftentimes this may wound the delicate egos of insecure field managers. "Nobody talks to *my* men without going through me." We've all seen these egomaniacs from time to time. Every organization has its ration of them. This is one of the problems you must anticipate and on which you must early get an understanding with top management. There isn't time to play games with egomaniacs after the RFP comes out. And if you still run into this problem, be prepared to go to the executive suite and pound on some desks. As I've said, you have been assigned the responsibility; they have got to give you the tools to implement it.

Now, about those team leaders: After you have assigned them the responsibility and provided them the necessary guidance, given them all the information that is available, pointed them in the right direction, let them take it from there. Let them select their own subordinates, organize their own contributors, and provide their own guidance in their own way. Exercise surveillance, yes; be prepared to step in and pick up the pieces, yes. But make them do their own thinking. You can't do other people's thinking for them.

You must keep an especially close eye on them during the outline phase of the proposal. If you see they need more guidance, tutoring, supervision, give it to them. This is where your deputy's help can be invaluable. If you conclude they just can't hack it, replace them at once. You won't have time to keep propping them up later, because the next phase is absolutely vital to a successful proposal.

This is the phase where the team leader gets the first draft from his contributors, reviews it, and returns it with directions for improvement and correction. He may have to go through this iteration two or three or more times before an acceptable *first* draft is ready for submission to the proposal manager. Obviously there is no way the proposal manager can have the time to painstakingly go through two, three, four iterations from some 20 or 30 proposal contributors. The team leader's function in decentralizing this task is indispensable to writing a good proposal. Of course, he is not expected to turn in perfect work. That is the function of the deputy (professional) proposal manager. You can't expect him to take garbage and turn it into a gourmet meal in the available time frame.

The proposal manager and his deputy should have the right to expect at least the following from their team leaders:

a. Input that follows the general outline
b. Input in the prescribed format
c. Input that addresses all applicable requirements in the SOW
d. Input written and organized in a logical, coherent manner

The proposal management team can then concentrate on improving the writing style, and can suggest improvements in technical approach, point out deficiencies, compare with RFP evaluation criteria, weave in applicable themes, check for thoroughness, and accordingly write a critique.

ESTABLISH *MODUS OPERANDI* CONTROL AND FEEDBACK SYSTEMS TO ESTABLISH TIMELINESS, QUALITY, AND EARLY IDENTIFICATION OF PROBLEM AREAS

As stated previously, a milestone chart should be prepared immediately after RFP release and disseminated to the troops. A large poster-size copy of this chart should be posted in the proposal area. Every effort should be made constantly to warn the proposal team that milestones must be faithfully met. Even so, there will always be some individuals who are late (but you must not condone it). For that reason it is best to leave yourself some cushion at the end of the schedule (disguised as printing and binding time, for instance) to pick up all the loose ends.

The *modus operandi* is set out, of course, in the proposal plan. It should also include a work flow chart similar to the one presented in Figure 6-4.

On this chart, the deputy proposal manager is assumed to be the corporate or local expert on proposal preparation. Note that he reviews and critiques each input only after it has been reviewed by the team leader and the proposal manager. Note also that the Proposal Team review and the corporate management review are accomplished concurrently. This saves time in that all changes, corrections, and revisions can be accomplished in parallel and not in series. Note also that it is a *Proposal Team* review. This is the best way to make a final review of the proposal. All key members gather around a conference table, each with a copy of the proposal that they have first studied and made notes thereon. Then the team goes through the proposal, page by page, incorporating changes as they go. Thus everyone has a chance to recommend corrections or revisions as he sees them, but the person who is responsible for each of the respective sections also has a real-time opportunity to defend his

FIGURE 6-4 Work Flow Chart

Review and correction will proceed
in the following sequence:

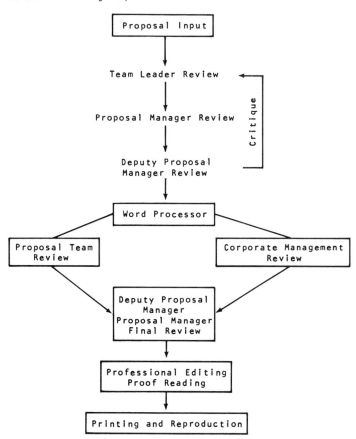

version if he so desires. This is an ideal way to avoid contradictions or inconsistencies in the proposal. It is probably also the best way to ensure that all RFP items have been addressed. It has been my observation that this is a more effective method of reviewing the proposal than the "Red Team" approach.

Finally, any editing changes resulting from the management reviews can then be incorporated by the professional editor before he releases the proposal to final draft preparation. Note that I have used the words "professional editing" as distinguished from amateur editing, that compulsive malignant disease that seems to inflict all senior managers. An executive salary does not automatically confer expertise in English. An education in English and years of experience in technical writing do. If only they would leave the editing and word smithing to these experts!

Generally, it is best to put only the first *corrected* drafts (that's the drafts that have been reviewed, corrected, and approved by the team leaders) on the word processor, because often there will be major revisions required of the preliminary rough drafts. Certain "boiler plate" sections, however, such as résumés, company experience, company administrative procedures, and so forth, can usually go directly on the word processor.

Review and correction at every step is to proceed expeditiously. This requires that it proceed in piecemeal fashion; that is, with various coherent segments being submitted and started through the cycle without waiting for an entire chapter to be completed before starting the review process. This is necessary to avoid a log jam of work to be done at the last minute, vitiating the effectiveness of the review function.

Some people I've worked with evidently think that a proposal effort is supposed to culminate in a kind of masochistic, herculean, cataclysmic endurance contest. The idea is, this is an opportunity for certain people to prove their devotion to duty by working 18 hours a day. On the contrary, it only proves the stupidity of some management teams and proposal managers for not organizing and budgeting the time available, making timely preliminary preparations, devising a schedule that people can live with, and ensuring that everyone abides by that schedule.

The trouble is in most outfits I know: management spends an inordinate amount of time enjoying the problem; that is, holding interminable meetings without ever determining any action items, bewailing the fact that the corporate office wasn't providing more help, wondering if they really ought to make a Bid decision, casting around for possible teaming arrangements, agonizing over the cost of writing the proposal, whining about all the difficulties involved, and on and on. I mean, *after the RFP has been issued*! Then the meeting breaks up and the only decision made was to meet again next week. No wonder people hate to get dragged into a proposal effort if management wastes half the available time just making decisions they should have made *before* the RFP came out. I had the miserable experience once of working on a *billion-dollar* proposal where management hadn't even made the Bid decision until *two months* after the RFP came out!

ESTABLISH INTERFACES REQUIRED FOR COORDINATION

Not the least of your challenges as a proposal manager is that of establishing useful and effective interfaces with a variety of management personnel whose interests are often in conflict with yours. At best, you should strive for enthusiastic participation from all concerned, but this is too much to expect. Nevertheless, you must demand at least grudging cooperation from all concerned. The key, of course, is top management, and you must exercise all your wiles, charm, and persuasiveness in getting strong, resolute, and *visible* backing from corporate management. If you succeed well in this, the other interface problems will usually take care of themselves.

Corporate Management

These people are looking only at bottom lines.

> What is it going to do to our budget?
>
> What is the potential profit?
>
> What is the benefit to the company, especially me, the manager/vice president/president?

The best way to get to these people is to convince them of the benefits inuring to them *personally*, if this proposal is a success. They will become heroes in the corporation for winning this new business that bodes well for future growth and profit for the company. This will be a giant leap forward for their careers.

Controller/Vice President, Finance

What you must bear in mind when dealing with these people is the fact that their objectives and philosophy are the very antithesis of that of engineers. Engineers are driven by the practical feasibility and realities of getting the job done and hang the cost. Cost factors are merely extraneous impediments. The driving factors (sometimes obsessive factors) are feasibility, technical approach, systems concept. The finance people (the bean counters), on the other hand, are grandiloquently unconcerned with such mundane technicalities. Their only concern is the bottom line—profit.

There is nothing wrong with this, of course. If engineers thought like accountants, they would not last long as engineers. If accountants thought like engineers, they would not last long as accountants. Accountants have to worry about the next quarterly report, and hang the technical details.

Just remember, when you talk to accountants, you must talk in language they understand: minimizing of risks; long-range versus short-range profit potential; anticipated average cost-per-man-hour; burden rates; magnitude of other direct costs (ODC); fee structure, G&A, and cash flow. Of course, you don't have all the answers, but talk about them and present an open-minded, pragmatic impression of "we are all in this together." Don't try to nail down any concrete agreements on anything in the early stages of the proposal. The time for that is when you have reached the point of no return in the proposal process. The bean counters will be more amenable to compromises on costs, margin of profit, and all that when it's too late to turn back, and especially if you have secured firm corporate backing.

Field Managers

These people are the most intransigent and obstreperous of all you will have to deal with. To characterize their usual attitude as parochial would be an understatement. One will occasionally find shining exceptions, of course, but by and large, they are overtly (sometimes violently) hostile to any "outsider" borrowing any of their fair-haired boys to help on a proposal. Their overriding concern is not the next quarterly report, but tomorrow's operation. The truth of the matter is nine times out of ten (and take it from someone who has been there) they have a nice comfortable routine, often have a fat-cat operation going, complete with coffee breaks, random bull sessions, time off to handle personal business, compensatory time off for professional-level people, and all the other perquisites that go with minimal upper management surveillance. Naturally they don't want anyone rocking the boat or forcing them to demonstrate to the world that their operation *can* function with one less person for a limited time.

The way to deal with these people is to try to persuade them that "we are all working for the same company," that company growth benefits them by providing greater job security and provides more opportunities for advancement in the company. If this doesn't work, remind them that you have the strong backing of top corporate management, which has promised you enthusiastic participation and full support in making this proposal a success. This latter approach often works miracles. Sloth is suddenly transformed into zeal, malingering into eager participation, lethargy into earnest desire. Sometimes there is no holding them back at this point. Your only problem now is how to harness their energy.

Your Current Customer(s)

Your customers' attitude is that they couldn't care less as long as your proposal effort does not impinge on their own contract with you. Keep a low profile, but be open and candid in answering any questions they may ask. And, above all, keep your proposal effort completely isolated from your current contracts—separate office facilities, absolutely no proposal work done on *their* time, and so forth. It wouldn't hurt to let your customer know that expanding your business base could eventually result in a lower G&A for you and thus lower contract costs for him.

The Supporting Cast

I previously alluded to the duties of the proposal coordinator in arranging for office space, typing, security protection, printing, editing, work flow, and so forth. There are too many interfaces involved here for you to deal with on a day-to-day basis. That is why you selected the proposal coordinator. Help him with the initial interfaces (pre-RFP) and see that he gets started off on the right foot, then leave the rest to him. After that you should have to interface only with him.

CONTROL AND FEEDBACK SYSTEMS

Many of the mechanics of proposal preparation that I have presented up to this point are designed to enable you to establish and exercise control and provide the means for early identification of problem areas with your proposal effort. Now let us examine some of the devices you must use to implement control and feedback systems. Without faithful attention to these devices you are headed for a very traumatic period of confusion and turmoil.

Device Number 1: The Proposal Plan

You must get this out before the RFP, based on what you know. Lay out your valid assumptions, set out preliminary instructions, let the key people in on your plans and what little you do know. They have to make plans too.

Device Number 2: The General Outline

It is most desirable that you get this outline out before the RFP, based on what you know. It is absolutely imperative that you get it out to all involved in the proposal as soon as possible after RFP release. This is the road map of your proposal. You can't expect people to start out without a road map.

Device Number 3: The Overall Organization

You must draw up a tentative overall organization chart for the contract during the research and analysis phase of your pre-RFP preparations for the proposal. Never mind if there are uncertainties at this point. This is a tool for helping you get a handle on the problem. This is a starting point for developing your whole proposal concept. This overall organization chart should be included in your proposal plan and updated as you accumulate additional information. See that it gets wide dissemination within your proposal team. They may have some ideas for improving on it. In any case, the overall organization chart is the foundation for the proposal that is to be constructed. You can't expect people to build without a foundation.

Device Number 4: The Proposal Schedule

See that this gets wide distribution. Post a large copy on the wall and have someone (your proposal coordinator) check off progress as you go.

Each of the team leaders should make up his own schedule to control the activities of his contributors. Refer to the sample schedule provided earlier in this chapter. The team leaders have 20 days, for example, to get their first drafts from proposal contributors. They then have an additional ten days to review, correct, rework and get the finished first draft in to the proposal manager. How they budget this time is their business. Presumably they are grown men (or women) and have enough management sense to work out those details themselves. If you don't have this much confidence in them, then demand that they submit their detailed schedule to you for approval on D day + 1.

But above all, make it clear from the very outset that you can't and won't tolerate slippage of the schedule. Wield the top management club on this if you must, but *demand* it.

Device Number 5: Team Leaders

This is a means of decentralizing your control into manageable segments. If it's a ten-man-year effort you are bidding, you don't need them, of course. But if it's a 50- or 100- or 1000-man-year effort, involving a multitude of disciplines, you obviously must break the total job down into bite-size chunks, unless you are Superman. This is basic, but it's funny how I have so much trouble convincing management that it is necessary. That's because a lot of management people do think they are Superman.

On the other hand, it's probably because most people just don't realize how much detailed work goes into writing a good, well-organized, polished proposal. The instructing, guiding, planning, brainstorming, organizing, writing, correcting, reworking, revising, rewriting, recorrecting, editing, and so forth, take time; there are only so many hours in a day and one person just cannot do it all. Of course, you can eliminate the team leaders and the proposal coordinator if you want to save manpower and turn in a sloppy, ill-conceived mish-mash that will be a disgrace to your company and subject you to scorn and contempt in the business community for the foreseeable future.

Device Number 6: Personal Monitoring

Don't think you can sit back in your office and wait for the proposal to flow in to you, even after all the preliminary work and guidance you have already provided. You must keep your finger on the pulse, so to speak. Seek out your team leaders; find out what problems they might be having; make sure everybody is proceeding in the right direction; and redirect their efforts, if necessary. And see that everyone is moving along in accordance with your schedule.

You may have to replace certain members of the proposal team. It happens all the time. Some people just can't hack it. The best time to do it is before the RFP comes out. The longer you wait, the more impossible the situation gets. That is why you must stay on top of every detail from beginning to end. If you don't, you may lose control, and if you lose control, you're dead.

Device Number 7: Work Flow System

A work flow diagram was shown earlier in this chapter. It is somewhat simplified for the sake of brevity. What is not shown is the function of the proposal coordinator in making this whole system work. You need to get together with him early in the project and work out a *modus operandi* for maintaining control of the paper flow. By the way, the paper flow can be a horrendous mess on any large proposal, that is, say, involving a contract over $50 million. And this is another factor that forces you to decentralize by using team leaders. That keeps the paper flow from becoming a problem, at least until you get a coherent and reasonably acceptable first draft.

Now the proposal coordinator must keep two loose-leaf notebooks, one for the first draft and another for the final (second) draft. Nothing goes into either notebook without specific approval of the proposal manager or his deputy. Nobody puts into or takes out anything from either notebook except the proposal coordinator. He is sole keeper of the notebook. That way, the proposal management team can keep track of the status of the proposal at any given moment in time. If one finds a blank where the computer maintenance section is supposed to be in the first-draft notebook, he knows the first draft is not finished yet. If he does find an input there, he can look it over and then check the second-draft notebook to see if the second draft has been completed. When the second-draft notebook has been filled up, the proposal is finished, ready for final editing and publication.

In summary, control involves:

- Feedback systems
- Personal monitoring
- Rigorous scheduling
- Review and correction to ensure quality
- Early identification of problem areas
- A system for controlling the paper flow

CHAPTER SEVEN

SOME IMPORTANT BUT OFTEN NEGLECTED ITEMS

I was going to start this chapter with a section on unsolicited proposals, but after much soul searching and an agonizing review of my own experience, I came to the conclusion that the best advice I can give you on this subject is, "Don't do it." That is, don't do it unless the "unsolicited proposal" you are considering has actually been solicited (unofficially, of course) by someone in a position of authority to accept your proposal and make an award. Otherwise, you are wasting your time and money. If the customer thinks you have a good idea there, he will simply take your idea and put it out on the street to see if he can get a lower bid than you are offering. That is one experience I had. The last experience I had developed as a result of a suggestion by a person *not in a position of authority* that we write an "unsolicited proposal" to do so and so. Well, we thought, "maybe he is delivering us a message from someone in authority." Many thousands of dollars later we gave them an unsolicited proposal. They dutifully reviewed it (I think) and summarily rejected it. Why? Who knows? When we pressed them for an answer, they politely told us they didn't ask for the proposal, so they didn't have to give us any answer. And they were right. They didn't, so they didn't, and that's the way it stands to this day. We paid our tuition on unsolicited proposals and thanks a lot. So instead of unsolicited proposals, this chapter begins with a few words of wisdom on consultants.

CONSULTANTS

The consultant is bound to rear his ugly head sooner or later in the field of marketing and proposal preparation. Some consultants (or perhaps I should say many) have given the consulting business a bad name and not altogether without good reason. They are known in some circles by such pejorative terms as "Beltway bandits" (in the Washington, D.C., area), "double dippers" (because so many of them retire from government service and then sell their expertise *and their connections* to the contractors), and there are other even less respectable terms.

You might say that consultants are, however, a necessary evil. They can be used in a variety of ways to improve your proposal. They can be of immense help in providing inside information (intelligence) on the target contract, or they can open doors to enable you to glean information yourself, or they can be extremely useful in reviewing and critiquing your proposal, or they can mess up your whole proposal and thoroughly confuse and demoralize your proposal team. It all depends on how you choose and use them.

The trouble with consultants usually isn't so much the consultants themselves as it is the careless lack of guidance they often have. That is, the consultant may be told simply, "Go down there and help out those guys on the proposal; they probably need all the help they can get." I have even seen management people who think you can actually hire a consultant to write your proposal for you. If a company cannot write the proposal themselves, how on earth do they expect to perform the contract? I knew a company once that hired a consultant to write most of their proposal. They tied with another company as losers in the competition. Funny thing, though, they found out about a year later that the same guy wrote the other losing proposal too. This is the sort of thing that gives consultants a bad name.

The proposal manager must exercise complete control of consultants, including veto power over any of their recommendations. He must be the one to decide, in fact, whether a consultant is needed at all, and if so, who is qualified to provide this assistance. He must establish an agreement with management that the consultant reports only to him, and for a specified purpose and for a specified time, and *always* on an exclusive contract. Before the arrival of the consultant, the proposal manager should draw up a detailed plan for utilization of the consultant, the parameters and objectives of his task, and so on.

First, the use of consultants should be limited strictly to their particular field of expertise. Second, the parameters of their consultation should be carefully and precisely defined. Third, their time should be carefully structured so that there is no wasted motion—no aimless bull sessions, no irrelevant recommendations. Remember, their time is expensive. And fourth, they should be given a statement in writing that stipulates exactly what is expected of them. I'll tell you an experience I had which aptly illustrates the above.

We were working on a large-scale multimillion-dollar contract proposal and our experts were hopelessly divided on a very important decision: (1) What should be the proper staffing level of one of the major facilities involved in the contract? and (2) Should we propose that the facility be relocated? Management negotiated with a brilliant young Ph.D. who had inside knowledge of the operation and a wealth of experience in this particular field. Trouble is, nobody gave him any guidelines as to what was expected of him. I have nothing but respect for Ph.D's *so long as they stay within their own area of expertise.* But many, it seems especially those who are in the consulting business (1) have delusions of grandeur insofar as any limitations to their intellect and (2) soon lose touch with reality when they stray from their own particular specialty.

This Ph.D. was in business for himself as a consultant so it was to his interest that he make as big a job out of this as possible. When he reported to me for work (I was the proposal manager), he immediately launched into a discussion of everything that could possibly be involved in the proposal, even our format for résumés, the executive summary, our company-related experience, qualifications of some of our key personnel, blah, blah, blah, yak, yak, yak. Not wanting to make a scene in front of the proposal team, I suggested we go into my office and close the door. I sat him down and said, "Now all I want from you is the answer to two questions, together with the rationale therefor. *And I don't want anything else!*" With that

I pounded my fist on the desk so hard the ash trays almost fell into his lap. Having thus gotten his attention, I told him I would give him a statement of his mission in writing the next day. Fortunately, I knew of his proclivity for enlarging on the original intent of the consultation and of his propensity for coming up with some asinine ideas once he got outside his particular field. Fortunately, management backed me up or a lot of money would have been wasted, and the proposal team would have probably been driven to distraction before he was through. Which brings me to the second thing to bear in mind about consultants: Make use of them as early as possible. *Do not* wait until the proposal is virtually finished and then have them come in and tell you what you did wrong when it's too late to make changes. (This same advice goes for management reviews.)

I hope I am getting this in the proper perspective, because I am not against using consultants, *per se.* The point I am trying to make is that, like any other proposal activity, they must be controlled, and they do have limitations. Oftentimes there is just no way you can learn enough about the details of an operation to write a winning proposal without getting hold of someone who has worked there, knows the environment, the interfaces, the idiosyncracies of the customer, the customer's mode of doing things, and so on.

I once managed a proposal on a government contract that was very complicated as well as being unique in many ways. Fortunately, my management secured the services of a recently retired military officer who was a key individual in this operation. I spent two entire days picking his brains and taking copious notes, and I'm sure we could not have won this contract without the information he supplied. But, knowledgeable as he was, he didn't know *anything* about the incumbent contractor's wage structure, the distribution of the labor grades, the off-site operation, or the logistic support requirements. You see, even he had limitations. This kind of information we had to either deduce or get from other sources.

The "other sources" might be nonconsultants (and I'm not going to get into any details on this), could be former contractor employees, disgruntled current employees, vendors who supply the incumbent contractor, and other contractors at the same location. There are lots of games people can play here. There are also, as I pointed out in a previous chapter, many documentary sources from which you can deduce such things as costs, staffing, current performance, and so on. Consultants are not infallible. You would be well advised to gather all this data you can, independently, and then compare them with what your consultants tell you.

To sum up then, the best way to avoid trouble with consultants is:

1. Do not hire them at all unless you have a specific finite need for their services that cannot be provided in-house.

2. Structure their time so that they are fully utilized *in their own field*. If you need them to help write your proposal you should NOT be bidding on the job.

3. Bring them in at the earliest possible time. If it's for briefing purposes, then *before* the RFP is issued, if possible. If it's for critiquing your proposal, it should be to critique the first *corrected* draft, *not the final draft*.

4. Tell them only what they need to know and no more. Consultants have been known to work both sides of the street, ironclad exclusive contracts notwithstanding.

5. Don't accept everything they tell you on blind faith. Use other experts available to you to analyze the data they provide. "Calibrate" them by asking a few questions, the answers to which you do know, to see if they really are qualified. Sometimes the competition might even plant them to mislead the unwary, so weigh carefully everything they tell you.

RED TEAM PRO AND CON

"Fire them all!" some would say, but I don't advocate going that far. First, let me define Red Teams as the "experts" that corporate sends out to review and critique your proposal when your proposal is about finished (the usual case). I suppose there is a place for so-called Red Teams, but not in the world of technical services proposal writing. If you are, on the other hand, writing a proposal for a prototype of an experimental airborne "widget," fine. Here you have something that will either fly or it won't fly, and its success is subject only to the inexorable and immutable laws of physics. So you call in some experts to question every assumption, equation, concept, all experimental data, mathematical approaches, in order to determine if there is any reason it won't fly. You are applying the scientific method here, and virtually nothing is left to subjective interpretation.

Just the opposite is true in the case of a technical services or hardware production proposal. There is never just one right solution to a problem. The best solution is usually a matter of judgment and taste, subject to the vagaries and idiosyncrasies of the customer. The correct solution is arrived at through a subjective process; the scientific method does not apply. Why am I telling you all this? Because here is what usually happens when you use this naive, juvenile, Red-Team approach in preparing a proposal for technical services. After weeks of 12- to 14-hour days (assuming the usual screwed up lack of prior preparation), you have finished the final draft of the proposal. You and everyone else on the proposal team are exhausted but exultant, confident that your efforts have been well rewarded by the satisfaction of a job well done. You have carefully analyzed the RFP, brainstormed the difficult places to death, arrived at a workable consensus on the complex decisions that were required, and reviewed and rewritten the drafts under the expert guidance of the proposal manager.

Then, along come these neophytes. They have not been in on the brainstorming sessions, are not privy to any of the logic that went into decisions, never even saw the RFP until last week. Now these people are going to tell you all the things you did wrong, on the basis of a few hours of casual review. (I use the word "casual" because they have no personal stake in the proposal.) Worst of all, in most cases you will get just as many conflicting solutions to any problem as you have members on the Red Team. You will get more gratuitous advice than you can absorb, and by the time they get through, you will be totally confused, frustrated, and discouraged.

In the first place, if an organization has someone on its staff who is smart enough to perform a critical and constructive review of the proposal, why not bring him in early to provide advice and counsel so the proposal can be done right the first time instead of after it is finished and then do it over again? The best procedure is to start review procedures at the completion of the first draft. Then after everything is proceeding in the right direction, the Red Team can fold its tents and quietly steal away forever.

In any event, the term Red Team should be abolished. It has the unpalatable connotation of an adversary relationship. Such a relationship is inimical to the process of writing good proposals. The proper connotation for management to project is the relationship of teamwork.

Don't leap to the conclusion that I am dogmatically out to abolish devil's advocates. Far from it. It's just that this sort of thing should *not* be done on an adversarial basis; it is simply counterproductive. By all means, every proposal needs critical review (and right now I'm talking about review in the technical sense), but such critical review should be performed in accordance with a set of common-sense rules. Here they are:

- The reviewers should have thoroughly studied the RFP.

- The reviewers should have in-depth training and experience in the field.

- The reviewers should have some experience in proposal preparation.

- The reviewers should be dedicated to the job, not dragged in against their will.

- They should be brought in early, after first doing their homework, usually after preparation of the first corrected draft, *not* after completion of the final draft.

- If extensive revision is indicated by the reviewers, they should present their critique in writing. Oftentimes the reviewers themselves will learn something they didn't know in the course of setting their thoughts down on paper.

- Reviews or critiques, even when reduced to writing, should be presented in person; or at least the reviewer must be made available for cross-examination.

- Criticism should be positive and constructive. If the reviewers believe something is wrong, they should have a practical suggestion for making it right. If the reviewers can't think of any remedies, they shouldn't be there.

- Sarcastic, degrading, or insensitive remarks should not be tolerated. The net effect can only be counterproductive.

- The proposal manager should have total veto power over the reviewers, as well as authority to remove same from premises.

I know some people will take issue with the last item, especially corporate management, because they are often the ones who send in these reviewers (qualified or not) to torment the proposal manager. One might ask, "If they don't trust the proposal manager to do the job, why didn't they pick someone else to manage it? If the reviewer is so smart, why didn't they have *him* write the proposal?" If they didn't pick a competent proposal manager, then they themselves failed in their duty as managers. Their mistake cannot be remedied by sending in a Red Team, because in the final analysis the success or failure of the proposal is going to be pinned on the proposal manager. He therefore absolutely has to have the authority to make decisions commensurate with his responsibility. Furthermore, corporate management should limit the exercise of its wisdom to the field of corporate management—decisions involving profit margins, tax matters, customer relations, corporate objectives, and so on. That will be the subject of the next section.

MANAGEMENT REVIEW—A NECESSARY EVIL

The general tenor of what I said about Red Teams applies as well to management reviews. The proper function of management is to manage. Top-level management should therefore be concerned with top-level management concerns: how to make money, how to manage corporate resources, how to sustain systematic corporate growth, how to develop new markets, how to promote the corporate image, and so on, and so forth. It follows that management reviews should be concerned with those issues that directly involve management: pricing strategy, availability of facilities and personnel to support the contract, financial risk involved, "political" considerations, the extent of corporate support required, and general conformity of the proposal to corporate standards and objectives. It should not involve wordsmithing such as changing "guidance" to "guide-lines," "use" to "utilize," "in case of" to "in the event that"; crossing out articles and conjunctions, sticking in exclamation points and semicolons and all that nonsense.

Lincoln was fortunate that he did not have a "manage-ment review" of his Gettysburg Address. Instead of "Four-score and seven years ago our fathers brought forth on this continent, a new nation conceived in liberty and dedicated to the proposition that all men are created equal," it would have turned out, "Approximately 87 years ago our forepersons initiated affirmative action to establish a new sovereignty, its genesis being fundamentalized in the premise that all persons are *ipso facto*, integral units of an egalitarian society." And so on, *ad nauseam* and thus, instead of an inspiring message for the ages, his words would have been lost forever on the evanescent air.

Management reviews should not involve nit-picking the grammar, the format, the mechanical details, the syntax, the spelling, and on and on *ad absurdum*. People with common sense judgment would be astounded at the number of high-priced executives that spend their time trying to second-guess a 22-year-old technical writer and actually think they are fulfilling a useful function thereby.

Generally, the practice is to wait until the proposal has otherwise been finalized before submitting it for management review—in other words, not until it is ready to go to the printer. This is wrong. It should be submitted as soon as the second draft is complete; *before* the editor has done his thing in preparing it for the printer, and *during* the period that the proposal manager and his key assistants are making their final review.

This method has the following advantages:

- It saves time. The proposal manager's review and the management review are being conducted concurrently instead of in series.
- Any extensive revision required as a result of management review can be accomplished with much less wasted motion at this time.
- Since it is impossible to prevent top management from playing editor, it eliminates the necessity of nit-picking changes after the professional editing has been accomplished.
- It gives the proposal team an opportunity to argue the advisability of management changes before they (the professional team) have reached the point of exhaustion and resignation.

Summary: Consultants and Reviews

Consultants

1. Use sparingly and for specific, structured, limited, documented purposes.
2. Limit use strictly to their field of expertise.

Review Teams (the so-called Red Teams or Tiger Teams)

1. If on an adversarial basis—forget it.
2. If to review the second draft, OK, providing:
   - they have thoroughly studied the RFP;
   - they have in-depth training and experience in the field;
   - they have extensive experience in proposal preparation;
   - they are dedicated to the job;
   - they present their critique in person *as a team.* "As a team" means a consensus.

If the above conditions are not met, review teams are counter-productive.

Management Reviews

1. Should be confined to management concerns.
2. Should be performed concurrently with the proposal team review.

THE BIDDER'S CONFERENCE

The bidder's conference is usually held a week or two after issuance of the RFP. This gives everyone a chance to become familiar with the RFP and hopefully ask intelligent questions, help the Government correct their typos, inadvertent omissions, and other careless errors. (There are always errors.) Or, in the case of a commercial contract for hardware, the customer can explain and clarify the design criteria. In the case of service-type contracts where the work is to be performed on Government premises (GOCO contracts: Government owned, contractor operated), the conference is held at the Government installation. This provides the contractor with a splendid opportunity to familiarize himself with the operation and its environment.

The trouble with some companies is that their management uses the bidder's conference as a chance for an ego trip or a junket to escape the humdrum of the office. They will send the vice president for marketing, the vice president for finance, and maybe, if he is lucky, the proposal manager. They will all wander around like a gaggle of tourists, renewing old acquaintanceships, buying drinks all around at the motel, and will generally have a helluva good time. Any information they may gather that is of use to the bedeviled technical proposal writer would be purely accidental.

The number of people permitted to attend the bidder's conference is nearly always limited to two or four. They should be chosen in the following order:

1. The proposal manager (provided he understands the contract, and provided he has a competent deputy to mind the store while he is gone)

2. The most knowledgeable technical person available (provided he knows how to take notes)

3. Someone who is good at counting (noses and dollars); a bean counter

4. Your second most knowledgeable technical person

These people should look upon the bidder's conference as a learning experience. They should be prepared to write a detailed report of the entire operation in all its facets upon their return and be able to answer any questions that may come up in the future.

They should be good at taking notes and remembering everything they see, such as the configuration of equipment, floor-plan layout, numbers of people working in the area. This is definitely not a reliable criterion, though. The incumbent will almost always try to confuse you. I remember one case where he had a three-shift operation and he kept the outgoing shift and incoming shift both on the job (working like crazy) until the tour was over to try to fool us into thinking he had twice as many people on the job. A better way to get an estimate of numbers of people is to count the numbers of work stations, look for duty rosters, items on the bulletin board, names on desks, work schedules, and other things that the incumbent forgot were posted for all to see.

Those making the tour should check with all proposal team members before leaving the office, to determine any specific bits of information they need to know. Sometimes just the manufacturer's name and model number on a piece of equipment can be of great significance. Most RFPs include a listing of all furnished equipment, but the listing may not tell you all you need to know. Also, the manner in which certain equipment is being put to use may be of vital importance in your concept of the operation or your staffing levels.

A good example of this is a proposal I once did for a contract in the Caribbean area. It included O&M of three Askanias (optical tracking mounts). These particular mounts were normally operated by two technicians, one for tracking in azimuth, the other for tracking in elevation. If tracking dynamic targets (like a missile or an airplane), these have to be highly skilled and experienced at optical tracking. We found out, however, that these mounts were used only for positioning of ships by triangulation—not exactly what you would call a dynamic target. So we bid one experienced technician, skilled in maintenance of the equipment, and proposed using the facility maintenance man (a kind of glorified janitor) to help out on the infrequent occasions when the mount was used for positioning a ship. We saved three man-years in our bid.

Sometimes the customer will allow cameras and tape recorders during the bidder's conference. If they didn't say you cannot, assume you can, but keep a low profile. It is generally frowned upon, and besides most people get very uptight when you stick a tape-recorder microphone in front of them. You are less likely to get an open, uninhibited answer to your questions or as detailed an explanation if you go around ostentatiously flashing that stupid tape recorder. In situations where taping and photography are not specifically prohibited, there are miniature tape recorders with microcassettes (good for an hour of taping), which can be carried in your jacket pocket and will record anything you can hear. There are miniature cameras with automatic exposure control, up to 36 frames per roll, which can fit in your vest pocket. I mention these modern devices as an alternative to those who are not very good at shorthand or sketching. According to the news media, there is something inexplicably sinister about the use of any device that replaces the old-fashioned means (a photographic memory and instant recall). However, for ordinary people like me, I recommend their use whenever permissible.

One other good use for the camera and tape recorder: There is often too much material concerning the contract for the customer to disseminate to all bidders. So he will set up a library of all pertinent information. Things like SOPs, maintenance procedures, workload data, schematics, floor plans, maps, training material schedules, and on and on. It is impractical for him to allow people to remove it from the library and the customer may feel (rightfully) indisposed to provide copying facilities. I say rightfully, because there will always be some inconsiderate hog who will want to copy everything in the library.

What you need to do is send somebody that is smart enough to sift out what is important and quietly read these items into the tape recorder. Or, if it is an important drawing or flow chart, take a picture of it. Saves a lot of time, and your good manners and consideration for others in minimizing the time you need the document will be appreciated.

Of course all this gathering of information will go for naught if it is not properly disseminated in a timely manner. You wouldn't believe how messed up some proposal efforts get when they are not competently managed. I once worked on one where two people attended a bidder's conference and tour for four days, equipped with tape recorders and cameras. But since the proposal manager didn't think it was important, the tapes (six hours' worth) were never transcribed and no debriefing was ever held. Vital workload data from the customer's technical library read onto tapes was never divulged to the proposal team, and in fact, no one on the proposal team received any useful data whatever from eight man-days of briefings and tours.

So if you are managing a proposal, see that the proposal team gets a debriefing, the tapes get transcribed, and any pictorial data are presented. Don't expect the proposal team to listen to six hours of tapes. It is your duty to see that all the extraneous material is removed after they are transcribed and present them a refined version. If you are not the proposal manager, then see that the proposal manager does it, even if you have to rattle someone's cage. What? You say it's too expensive to have someone transcribe those tapes? Then go back to the Bid/No Bid decision step. You probably can't afford to bid on this contract.

Some proposal managers are reluctant to ask any questions at the conference on the theory that the answers would benefit the competition. This theory has some merit, of course, but it has been my observation that this strategy is much overdone. If there is, for instance, a very real ambiguity in the RFP that cannot be resolved, are you just going to act stupid because Brand *X* might benefit from getting it resolved too? If there is an incumbent, you can be sure *he* is not going to be in the dark, so you will just bestow on the incumbent an advantage over yourself. Of course, if you are 95 percent sure you know the answer (perhaps because you have inside information that most competitors don't have), then don't ask the question. But, don't just assume you are smarter than the competition or that you are more clever. This pernicious delusion is guaranteed to lead to your downfall.

On the other hand, there may be occasions when you can ask a question for the sole purpose of confusing the competition. Let us say that you know the Government's engineering staff has spent a bundle of money researching the feasibility of using a laser-ranging device as part of their missile tracking equipment and have even budgeted funds for a prototype. But, you have a reliable friend who is a vendor of such equipment and has been intensively tracking the progress of this engineering study. He informs you confidentially that:

1. The system is perfectly feasible, but there is no one in the Government's Operations staff who even knows how to spell laser.

2. The chief civil servant involved in the Go/No Go procurement decision is a technical weakling and is scared to death of sticking his neck out to spend money on such state-of-the-art equipment that in his uninformed opinion might not work out. Besides, he is going to retire in a couple of years. (Why take chances?)

At the bidder's conference, you ask if the Government is going to release its engineering report on the laser-ranging system to the bidders. You don't care what the answer is, because you know the system will never become operational during the term of the contract. You also know the Government is *not* going to say, "Don't worry about the laser-ranging system. We just wasted $X00,000 of the taxpayers' money on that study." Instead, you will get the usual weasel-worded bureaucratic obfuscations. Now all the other bidders will waste several man-weeks of engineering effort feverishly studying laser-ranging devices and trying to find résumés of people with laser experience. They can't afford not to.

Another gambit I have seen tried successfully involves putting the Government on notice that you are not going to play their little games without protest. It is true that an official protest is almost never upheld, no matter how flagrantly the Government has violated the rules. But the threat of protest nevertheless can be a potent force in restraining them from cheating.

For example, suppose you have a SEATA contractor (systems engineering and technical assistance) whose function, among other things, was to help the Government write the RFP, and now this same contractor is eligible to bid on this RFP. (These things do happen.) This is a good time to get together with your fellow competitors (except the SEATA contractor) and see if you can reach a consensus that this so-called procurement competition is fixed, that the SEATA contractor has an unfair advantage and that the Government is just going through the motions of a competition to satisfy the letter if not the spirit of procurement regulations, having already decided they are going to give it to the SEATA contractor. Everybody gets up and fires off a series of pointed questions at the bidder's conference directed at the integrity of the competition, the methods involved, the SEATA contractor's relationship to the Government, and so on. Finally, after some 45 minutes of this badgering, the Government gets the idea that there is bound to be a protest here if the contract is awarded to the SEATA contractor, even if everything was done honestly.

True, the Government would win on the protest. They always do. But what an agonizing experience to go through! They have to answer questions from higher authority, from the inspector general, maybe even go before a Congressional committee. They have to prepare long, written reports, comb through their files to build up a case to justify their actions, research all the loopholes in the appropriate procurement regulations, even perhaps (horrors!) work on a weekend to get all their tracks covered in time to answer the protest.

Nobody wants to go through all this, so the prudent Government contracting officer will wisely see to it that the SEATA contractor does not get the contract.

There is nothing so demoralizing to a proposal manager as to go to a bidder's conference and see one of the various contractors present, casually and comfortably fraternizing with one (or more) of the Government procurement people. One inescapably gets the feeling one is on the outside looking in, and one hasn't got a chance of winning against these competitors who seem to be such good buddies of the Government contracting people. So if you have established a personal relationship with some of the Government people present, by all means greet them cordially as soon as the opportunity presents itself, and if it's during the coffee break, tell them the latest joke you've heard. You just might discourage some competitor enough for him to make a No Bid decision, and one less competitor is just as good as one defeated competitor.

In recent times there have been an abundance of what they call A-76 contract opportunities, so named because they stem from OMB Circular A-76, which directs government agencies to contract out government-operated (both civil service and uniformed personnel) functions, provided a cost savings to the government could be realized, and provided there is no overriding justification for the function to continue to be performed by civil service or military personnel. It all makes eminently good sense, of course, if one is to embrace the bold view that the function of government should be to govern, and not to operate photo labs, grocery stores, garbage trucks, furniture warehouses; or to try to design and manufacture optical tracking systems, shuttered video systems, and the like. Of course, these functions may require Government surveillance and overall direction. But, the functions do not have to be performed by an army of bureaucrats with all the civil service perquisites: 30 days' vacation after one year on the job, automatic promotions and pay raises regardless of performance, a benign retirement system, and red tape comparable to the passage of an act of Congress just to get someone fired for misfeasance, malfeasance, or even nonfeasance.

Successive administrations have thus dragged a multitude of bureaucrats kicking and screaming to the Government conference rooms across the land to preside over the dissolution of their cozy little empires. This is termed the "contracting out" process, the mere mention of which is bound to raise the hackles of almost any civil servant. By and large, the uniformed military services have gone happily along with the idea, adopting the pragmatic view that the business of the Army is soldiering, of the Air Force to keep 'em flying, and of the Navy to sail and shoot, not keeping detachments of sailors in places like Indian Wells or Navajo Springs making a daily count of the mothballed Amphtracks waiting in the desert air for the next war to start. The military forces realize that the more uniformed people they can place in combat and in direct combat support roles, the stronger our armed forces will be, and the more the taxpayer will get for his money. After all, that is why the taxpayer put all that money into training those men for the armed forces—to carry out *military* operations, not to shuffle papers in St. Louis or perform janitorial duties on some lonely post in the West that should have been sold to the local citizenry back when the Indian wars phased out.

Some time ago I attended a bidder's conference at one of those superfluous military installations. It involved an A-76 contract, in this case, conversion from civil service to contractor operated. Incidentally, this installation *had* been declared inactive by the military, and then the local politicians and congressmen got in the act and forced the military to keep it on the active list, though it had no justifiable excuse for its existence. They are probably the same congressmen who wail loud and long about the "irresponsibly huge Federal deficits" and the "unjustifiably high cost of our National Defense Establishment."

Back to the point I was about to make, which is that you have to read carefully the fine print in these A-76 RFPs. The bureaucrats often resort to all kinds of subterfuges to prove they (the government) can do the job more economically than contractors. That way they can circumvent A-76. In this case they slipped in one little sentence that said, "The contractor shall provide his own office furniture." On this contract, about 80 percent of the personnel would require desks, tables, bookcases, filing cabinets, and so on, so this would be a very substantial item in the bid figure. At the bidder's conference, one contractor asked if the government was going to replace the existing furniture and fund the resulting cost into their own government cost estimate. "No." "Then how can this be considered a realistic cost comparison between government and contractor costs?" Reply: (government weasel words). Another contractor: "What are you going to do with the furniture already there if a contractor takes over?" "We have other uses for it." Contractor: "The papers have been full of the scandalous situations of GSA warehouses overflowing with surplus furniture while the Government continues to buy more, and you are saying you need this furniture for other purposes?" Government: "Gentlemen, this meeting is adjourned. Any further questions will have to be submitted in writing."

That is the last I ever heard of this particular A-76 competition. Somebody told me the government decided they would have to revise the RFP. That should take at least two years while they figure out some other clever subterfuge to circumvent A-76.

DELIVERING THE PROPOSAL

I have really heard some Hairbreadth Harry stories on this subject. One of the best was the guy who arrived at the contracting office right on time, about a half hour before the deadline, having flown in from his home office 1000 miles away, only to discover he had forgotten to bring the cost proposal! The cost proposal was for a contract worth about $80 million. I don't think the guy even dared to go back to clean out his desk. He probably decided that 1000 miles between him and his boss was just about right.

There are many ways these days to deliver a proposal. If it is for a substantial contract, I would consider the U.S. mail service at the bottom of your options—with all due respect to our intrepid postal employees never allowing snow, sleet, or gloom of night to keep them from their appointed rounds. I know of a package that was addressed to the Marshall Islands in the mid-Pacific that ended up in Oslo, Norway. I sent a letter to Lagos, Nigeria, once and got it back six months later saying there was no such place. I mailed a letter to Alabama—"*AL* 35803." It ended up in Alaska.

For a small contract proposal that you could put in one or two boxes, I suppose you could trust a reputable air express company such as you see advertised on TV. If it's a large proposal, I think there is only one acceptable way: Deliver it yourself. If it is too bulky to carry on board as hand baggage, demand that you be allowed to follow it through the loading process so that you can actually observe it being loaded on board the aircraft. If the airline gets sticky about this, try another airline, explaining your reasons. Then, when you get off the plane, observe the *un*loading process. Make sure all boxes are well marked and sealed and that you count them coming and going. You should have someone meet you at the destination who knows exactly how to get from the airport to the contracts office. It's actually best to give yourself a 24-hour leeway if you have a safe place at the destination to store them overnight.

I once had a Hairbreadth Harry experience of my own. The proposal was delivered to me in Atlanta at the airport by a courier who had flown it in from Honolulu. I was supposed to catch the next plane to Huntsville, Alabama, where it was due at 2:30 P.M. No problem. My plane took off at 8:30 A.M. for the 45-minute flight to Huntsville. A slight problem developed, however. When we got to Huntsville, the airport was fogged in. No problem. We would land at Muscle Shoals and I could rent a car and drive back to Huntsville. Trouble is, Muscle Shoals was also fogged in. Back to Huntsville, where we circled for about 45 minutes; no dice. So back to Atlanta. Next scheduled flight— 2:30 P.M. No way. So I had a daredevil charter pilot fly me in a puddle hopper, just above the treetops and under the fog, to the Huntsville airport, jumped into a waiting car, and off to the contracting office with five minutes to spare. Next time ... Oh, I forgot to mention, we lost the contract to an outfit called GE, otherwise known among the aerospace community as Generous Electric.

One naturally asks, why wait until the last minute to deliver the proposal? For one thing, you just don't want to run the risk of someone in the contracting office being able to peek at your bottom line and relay this information to one of your competitors in time for him to make some last-minute changes to his cost proposal.

Another reason is, often the people delivering the proposal have to sign a log recording their visit to the contracting office, for whatever purpose. If you are the last person signing their log you can see who your competitors are. This may not help you much, but it's comforting to know, and besides, it might help you in developing your best and final offer (BAFO) strategy to know whom you are going to have to beat.

In the not-so-good old days there were a lot of shady characters playing some very unethical games, like bugging the competitor's proposal offices, planting spies, bribing printers, enlisting the help of janitors, and even breaking and entering. I talked to one unscrupulous character who said he always called up the airline and cancelled the reservations of his competitors on the day before they were to deliver their proposals. I said, "How would you like it if they did that to you?" He replied, "Never happen. I always make my reservations for myself, plus three fictitious names."

I think we have become a little more sophisticated and restrained in these contemporary times, but it still doesn't hurt to be too careful.

THE DEBRIEFING

After all the hard work and dollars you invested in this proposal, suppose, just suppose the unthinkable, the impossible does happen—you lose! As they say when you get a lemon, make some lemonade. Request that the customer provide you with the rationale for his decision. Ask him to tell you what you did wrong. Don't argue with the customer's logic even if it seems illogical to you. Make an honest effort to determine what you did wrong (or at least what they *say* you did wrong), so you can learn from your mistakes and do better next time. All the money and effort you invested in a proposal that does not win is totally wasted if you do not learn from the experience. If you learn enough to write a winning proposal next time, it is money well spent. You can't expect to win them all, so learn all you can from the ones you lose. Too many companies just drop the whole thing in bitter disappointment, go about making excuses for their failure, look for a scapegoat (usually the proposal manager; he is the handiest one around), and fail to investigate the real causes and document them for future use. Sometimes it can be a single thing like an error in your costing assumptions or your program manager was unacceptable to them or you inadvertently neglected to address a key requirement in the RFP.

If it was for a government contract, you have a right to demand a debriefing if the contract was for more than a stipulated amount of money. Some companies just send their marketing man around to these debriefings as a kind of perfunctory gesture. And he in turn, writes a brief perfunctory note to the vice president, or other person who perfunctorily reads it and throws it in the wastebasket. This is wrong, wrong. The marketeer seldom knows much about the technical details of the proposal and cares less. The proposal manager should go, and along with him, the sharpest person you had on the proposal team. Afterwards, they should write a memorandum for the record and distribute it to everyone on the proposal team. Also, place a copy in the files for the benefit of anyone who prepares a similar proposal in the future. The people on the proposal team will appreciate your honesty and thoughtfulness in giving them the facts and may also learn something in the process. They will also appreciate management's candor in explaining the mistakes. No one should be ashamed of making an honest mistake. Everyone should be ashamed of not learning from his mistakes.

THANK YOU

If there is only one lasting impression this book makes on pro-posal managers, and management in general, I hope it is here. Because, please, proposal managers, the people who provided concrete positive efforts and sacrifices to create the proposal—let them know you appreciate it. You may want them to help you again sometime, and this may be all the thanks they will ever get. I don't mean form letters to everyone that contributed to the proposal—the usual case. Take the time to write out personal letters (copy to his personnel file and top manage-ment) to each individual who *really* contributed to the proposal effort. As I have said before, it is easy (and inexpensive) to write an inept proposal. Any fool can do a cut-and-paste job and regurgitate the RFP. You don't have to thank these people. If you do, you will be cheapening the whole idea. And don't play politics with this. Acknowledge only the real contributors and you may be fortunate enough to gain the respect and loyalty of the people who really count in the long run. It only costs a postage stamp, and that may be the best investment you or your company ever made. And last, if you win the contract and your company will provide bonuses, see that they are awarded to the deserving—not the talkers, but the writers.

CONCLUSION: NOBODY PROMISED YOU A ROSE GARDEN

The instructions embodied in this book are based upon many years of practical experience involving a wide variety of pro-posals, both government and commercial. They say that "experience is the best teacher, but the tuition is often exorbi-tant." How true! Much of what has been presented here has been learned by trial and error and bitter disappointment. I hope that by applying the lessons learned here you can be spared the trial of "blood, sweat, toil, and tears" that I have endured in getting here. (I say blood *literally*, and I have the ulcers to prove it.)

If the instruction presented here is steadfastly applied by competent personnel, there is no legitimate way you can fail to have a technical proposal in the zone of consideration. Once you get that far, it becomes basically a matter of who is willing to perform the contract at the lowest cost. At least, that is the way it is supposed to work.

The technical proposal gets you in the door. After that, there are many subtle, hidden factors that determine whether you get the contract. While these factors may not be objective, you really can't blame the customer for finding ways to reject you on the basis of some of these factors. For example, suppose you show up at the Q&A session for an engineering contract looking like a bunch of hippies. You can't blame the customer for finding ways to eliminate you. You can't expect him to work with people whom he finds offensive. Suppose you show up, querulous, contentious, or unresponsive to his questions. Same thing. You will lose, and it will be perfectly legitimate and legal, too. You can count on it—regardless of how low your bid was or how good your proposal was.

Suppose the contracting officer's son-in-law, Joe Blow, has a key position with one of the competitors, or something else like that, if you know what I mean. That is something you can't do much about—except project the image of a very businesslike *no nonsense* outfit that won't hesitate to file a protest or a lawsuit if they are presented with a justifiable case.

The customers with whom you will be dealing normally have an ingenious and well-designed system to ensure honesty and objectivity in the procurement process. If faithfully implemented, the system would completely achieve its intended purpose. But you must bear in mind that no system that includes the human element is infallible. It works only as well as the quality and integrity of the people implementing it.

Struggling through a proposal, like struggling through life, is the art of compromise. The successful proposal manager must not only guide and inspire, but also demand and admonish. He should know there is a time to lead and a time to follow. The desirable must sometimes yield to the attainable, the theoretical to the practical, the hope to the reality. And sometimes too, the rules of fair play must succumb to the chicanery of the marketplace.

So now you have it all—the methodology and the reality. Good luck!

Index

DAVID A. WATTERS